T0138532

The Fire Ant Wars

The Fire Ant Wars

NATURE, SCIENCE, AND PUBLIC POLICY
IN TWENTIETH-CENTURY AMERICA

Joshua Blu Buhs

The University of Chicago Press Chicago and London

Joshua Blu Buhs received his doctorate in the history and sociology of science from the University of Pennsylvania in 2001. He currently lives in Japan.

The University of Chicago Press, Chicago 60637
The University of Chicago Press, Ltd., London
© 2004 by The University of Chicago
All rights reserved. Published 2004
Printed in the United States of America

13 12 11 10 09 08 07 06 05 04   1 2 3 4 5

ISBN: 0-226-07981-3 (cloth)
ISBN: 0-226-07982-1 (paper)

Library of Congress Cataloging-in-Publication Data

Buhs, Joshua Blu.
  The fire ant wars : nature, science, and public policy in twentieth-century America / Joshua Blu Buhs.
      p. cm.
  Includes bibliographical references and index.
  ISBN 0-226-07981-3 (cloth : alk. paper)—ISBN 0-226-07982-1 (pbk. : alk. paper)
  1. Fire ants—Control—Southern States—History—20th century. 2. Fire ants—
Control—Environmental aspects—Southern States—History—20th century.
  3. Insecticides—Environmental aspects—Southern States—History—20th century. I. Title.

  SB945.F535B84 2005
  632'.796—dc22   2004008495

⊗ The paper used in this publication meets the minimum requirements of the American National Standard for Information Sciences—Permanence of Paper for Printed Library Materials, ANSI Z39.48-1992.

The question of questions for mankind—the problem which underlies all others, and is more deeply interesting than any other—is the ascertainment of the place which Man occupies in nature and of his relations to the universe of things. Whence our race has come; what are the limits of our power over nature, and of nature's power over us; to what goal we are tending; are the problems which present themselves anew and with undiminished interest to every man born into the world.

Thomas Henry Huxley, *Man's Place in Nature*

Contents

Acknowledgments

Writing a book must be a little like raising a child, I think, although I have no children and this is my first book, so I can't be certain. *The Fire Ant Wars* began life in 1998 as an essay on *Solenopsis invicta* for Daniel Janzen's humans and the environment class at the University of Pennsylvania. It grew into my dissertation. Along the way, like any parent, I received copious amounts of help. Fellow graduate students in Penn's history and sociology of science department, among them Eve Buckley, Nathan Ensmenger, Carla Keirns, Erin McLeary, Susan Miller, Elizabeth Toon, and Audra Wolfe, offered advice and good cheer. Eric Richard and Vicki Tardiff, Dan and Debbie French, and Betty Smocovitis opened their homes to me while I was doing research in the field and I cannot hope to repay their generosity. They provided a warm reception and at least good-naturedly feigned interest when I rambled on about my research. Robert Kohler, Henrika Kuklick, and Susan Lindee served on my dissertation committee and, more important, provided models for what good scholarship should be.

As the project developed from essay to dissertation to book, it sometimes confounded me; it also surprised, delighted, and educated me. An author knows that he or she is never fully in control of a book: it has a life of its own and forces you in directions you never imagined. I could not have found the material that I needed without Peter Branum, Leonard Bruno (who somehow found the Creighton papers), Dwayne Cox, Janice Goldblum, Norwood Kerr, Becky Jordan, Steve Johnson, Kristen Nyitray, Sydney Van Nort, Robert Young, the interlibrary loan staff at the University of Pennsylvania, and archivists at Harvard, the Mississippi Department of Archives, the National Archives, and Yale. Having found the material, I could not have made sense of it without the help of J. Lloyd Abbot Jr., William Banks, Pete Daniel, John George, Linda Lear, Daniel Leedy, Clifford Lofgren, Robert Mount, Phil Pauly, Walter Rosene, Robert Rudd, Roy Snelling, Dan Speake, Jeffrey Stine, James Trager, Walter Tschinkel, Phil Ward, David Williams, and E. O. Wilson. I have never met Murray Blum in person, but over the course of my work on this project he became a friend. He offered constant encouragement and an unending stream of material. This book would be a lot poorer were it not for his aid. Three anonymous reviewers for the University of Chicago Press also helped to clarify my thinking and writing.

The University of Pennsylvania's School of Arts and Sciences, the National Science Foundation, and the National Endowment for the Humanities all helped to fund the work.

An excerpt of a letter from Rachel Carson to J. Lloyd Abbot is reprinted by permission of Frances Collin, Trustee, all rights reserved.

David Bemelmans, Christie Henry, Jennifer Howard, and the staff at the University of Chicago Press held my (virtual) hand through the long process of editing. They made the book better.

And then, after all that work, the project grows into a book and leaves. It will be odd not to have it around anymore. It has been my constant companion for six years now, from Pennsylvania to California to Japan. I have done as well as I can by this book, but, like parents everywhere, I'm bound to have made mistakes; I take full blame for those. I can only hope that this little book that I launch into the big world is strong enough to overcome those foibles and succeed on its merits.

Writing *The Fire Ant Wars* has given me a better appreciation for my own family, which too often I neglected in order to spend time writing or reading about insects. This book would not have been possible without the support I received from my parents, my brother, and my sisters. Above all, though, it would been inconceivable without my wife, Kim McIlnay, whose love, nourishment, and good humor has sometimes been all that has kept me going and has ever been all that I need. Kim, this book's for you.

Introduction

the aeolian
syrinx, that reed
in the throat of a bird,
when it comes to the shaping of
what we call consonants, is
too imprecise for consensus
about what it even seems to
be saying: is it *o-ka-lee*
or *con-ka-ree*, is it really *jug-jug*,
is it *cuckoo* for that matter?—
much less whether a bird's call
means anything in
particular, or at all.

Amy Clampitt, "Syrinx"

This book is about nature—the ways that Americans thought about it during the twentieth century, the ways we have transformed it, and the ways, in turn, that we have been changed by nature. It examines this relationship through the story of the red imported fire ant, *Solenopsis invicta*, a much celebrated and much loathed insect that plagues the South to this day. The ant arrived in North America during the 1930s or 1940s and spread slowly from its entrepôt in Mobile, Alabama. In the 1950s, it underwent what ecologists call an "irruption," which resulted in an explosive rise in its population. The ant spread over more than 20 million acres by the end of the decade. *Solenopsis invicta* built large, earthen mounds that interfered with tractors, decimated the native ant fauna, ate crops, attacked wildlife, worried domestic animals, and stung painfully, on rare occasions killing people (fig. 1). In 1957, the Plant Pest Control Division of the U.S. Department of Agriculture (USDA) initiated the first of two attempts to eradicate the insect using chemical insecticides. Over the next twenty-one years local, state, and federal governments spent more than $200 million to spray insecticides onto more than 57 million acres in hopes of killing the insect.[1] As the campaign progressed, though, it became clear that the insecticides also killed wildlife. Meanwhile, some questioned if the ant was really a pest at all. Opposition to the eradication

1. The cost of the program is from M. K. Hinkle, "Impact of the Imported Fire Ant Control Programs on Wildlife and the Quality of the Environment," in *Proceedings of the Symposium on the Imported Fire Ant*, ed. S. L. Battenfield (Washington, D.C.: Government Printing Office, 1982), 140. The number of acres sprayed is from R. L. Metcalf, "A Brief History of Chemical Control of the Imported Fire Ant," in *Proceedings of the Symposium on the Imported Fire Ant*, ed.

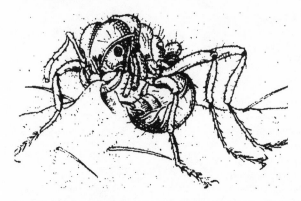

Figure 1. The imported fire ant as drawn by the USDA. This illustration shows the ant preparing to sting. It has grasped a hunk of human flesh and its gaster is bent so that its stinger can pierce the skin. The image was widely used in newspaper articles on the insect. (From "$10 Million Sought to Fight Fire Ants," *Montgomery Advertiser*, 8 March 1957, fire ant files, box 65, Entomology Research Division: General Correspondence, 1954–1958, 5 [UD], Agricultural Research Service papers.)

campaign mounted: the fire ant wars not only pitted humans against insect, but humans against humans. In 1962, Rachel Carson excoriated the fire ant campaign in *Silent Spring*. Two decades later, the Environmental Protection Agency (EPA) outlawed the chemicals that had been used to kill the insect. Throughout it all, the ant continued to spread, occupying 300 million acres by the end of the century and finding a place in southern culture—the star of books, songs, and festivals.

Entomologists, journalists, scientists, and historians have all chronicled the fire ant wars.[2] For the most part, these histories aim to explain why the USDA spent so much money and so many years trying to eradicate the insect. Why had the USDA engaged in such reckless and wasteful behavior? In my account, I also address this question and shed new light on the department's motives. During my research in the

S. L. Battenfield (Washington, D.C.: Government Printing Office, 1982), 124; C. S. Lofgren, "History of Imported Fire Ants in the United States," in *Fire Ants and Leaf-Cutting Ants: Biology and Management,* ed. C. S. Lofgren and R. K. Vander Meer (Boulder: Westview, 1986), 42.

2. For histories of the fire ant wars, see H. Wellford, *Sowing the Wind* (New York: Grossman Publishers, 1972), 286–309; P. M. Boffey, *The Brain Bank of America: An Inquiry into the Politics of Science* (New York: McGraw-Hill, 1975), 200–26; C. S. Lofgren, "History of Imported Fire Ants in the United States," in *Fire Ants and Leaf-Cutting Ants: Biology and Management,* ed. C. S. Lofgren and R. K. Vander Meer (Boulder: Westview, 1986), 36–47; P. Daniel, "A Rogue Bureaucracy: The USDA Fire Ant Campaign of the Late 1950s," *Agricultural History* 64 (1990): 99–114; E. F. Shores, "The Red Imported Fire Ant: Mythology and Public Policy," *Arkansas Historical Quarterly* 53 (1994): 320–39; P. Daniel, *Lost Revolutions: The South in the 1950s* (Chapel Hill: University of North Carolina Press, 2000), 78–87.

National Archives I came across 350 boxes of material from the Plant Pest Control Division's files—records that, so far as I know, no other historian has ever used. But I am also in pursuit of a different quarry: I want to learn what the fire ant wars have to tell us about the relationship between humans and nature.

Building on the work of environmental historians, especially Arthur McEvoy, William Cronon, and Richard White, I look at this relationship in three distinct but interrelated ways.[3] First, I consider nature to be something independent of humans. The ant's existence owes nothing to *Homo sapiens:* the insect is natural. Second, nature exists in our heads. We analyze and divide the world around us according to our preconceived notions, our place in society, and our experiences with the world.[4] Americans named the fire ant, we talk about it, and think about it. Finally, nature is something that we do. Nature existed long before humans, and will continue to exist if we disappear. But, while we are here, and as we grow in our capacity to dominate the environment, we increasingly shape the world around us. People brought the ant to the South, people altered the southern environment, and people made the chemicals used to kill the insect. Some of these things may seem unnatural: the ant is an import, not a natural resident of North America, and the insecticides that were used against it were synthesized, not products of nature. Both the ant and the chemicals, however, are composed of atoms, and while humans have learned how to move them around, we have done so according to natural laws. Invasive

---

3. A. F. McEvoy, "Toward an Interactive Theory of Nature and Culture: Ecology, Production, and Cognition in the California Fishing Industry," *Environmental Review* 11 (1987): 289–305; W. Cronon, "The Uses of Environmental History," *Environmental History Review* 17 (1993): 1–22; R. White, *The Organic Machine: The Remaking of the Columbia River* (New York: Hill and Wang, 1995). For overviews of environmental history, see D. Worster, "History as Natural History: An Essay on Theory and Method," *Pacific History Review* 53 (1984): 1–19; R. White, "American Environmental History: The Development of a New Historical Field," *Pacific History Review* 54 (1985): 297–335; W. Cronon, "Modes of Prophecy and Production: Placing Nature in History," *Journal of American History* 76 (1990): 1122–31; R. White, "Environmental History, Ecology, and Meaning," *Journal of American History* 76 (1990): 1111–16; D. Worster, "Transformations of the Earth: Toward an Agro-Ecological Perspective in History," *Journal of American History* 76 (1990): 1087–1106; W. Cronon, "A Place for Stories: Nature, History, and Narrative," *Journal of American History* 78 (1992): 1347–76; I. G. Simmonds, *Environmental History: A Concise Introduction* (Oxford: Blackwell, 1993); D. Worster, "Nature and the Disorder of History," *Environmental History Review* 18 (1994): 1–15; M. A. Stewart, "Environmental History: Profile of a Developing Field," *History Teacher* 31 (1998): 351–68.

4. The literature on the relationship between ideas about nature and the structure of society is vast. For introductions, see D. Bloor, *Knowledge and Social Imagery* (Boston: Routledge and Keegan Paul, 1976); H. Kuklick, "The Sociology of Knowledge: Retrospect and Prospect," *Annual Review of Sociology* 9 (1983): 287–310; J. Golinski, *Making Natural Knowledge: Constructivism and the History of Science* (New York: Cambridge University Press, 1998).

insects and poisonous chemicals may seem objectionable, but, in a meaningful sense, they are not unnatural: they are the nature that we produce.[5]

Chapter 1 offers a précis of this argument as I describe the ant's introduction into the United States. The insect evolved in South America millennia before the first human set foot on the planet. The ant is nature independent of humans. But people made the insect's transport to North America possible and Americans unwittingly facilitated the ant's spread. Adapted to life on a flood plain, the ant succeeded best in areas of ecological disturbance. In the third and fourth decades of the twentieth century, the American South underwent a huge change, what historian C. Vann Woodward calls the "bulldozer revolution." The ant, I argue, exploited this revolution to spread across the region. Thus it was a combination of the ant's natural history and human action that caused the insect's irruption.

Solenopsis invicta, though, also exists in people's imaginations, and in chapters 2 and 3 I examine how different groups have imagined the insect. I also try to account for these different perspectives. Warriors against the fire ant, I argue, looked at the ant through prisms shaped by preconceived notions, from a vantage point determined by their place in American society, and limited by the intellectual and cultural resources that society provided for them. When the fire ant first arrived, Americans saw it as a pest; as the insect spread, many came to believe it was among the worst insect scourges to ever invade the United States. "The government should be building as big a defense against the fire ants as they are against the Russians," an Alabama farmer said in 1957. "The ants have already invaded."[6] The USDA's Plant Pest Control Division answered these complaints by initiating a campaign to eradicate the ant. The plan was not to control the ant—reduce its population to an acceptably low level—but to eradicate it, to kill every single specimen of *Solenopsis invicta* within the United States. That decision, I suggest, was only partially a response to the problems caused by the ant. The Plant Pest Control Division (PPC) was a relatively new agency, as was the belief that chemical insecticides could be used to eradicate pests. By eradicating the ant, PPC officials could protect the South from an insect predator while both proving the validity of

---

5. For a fuller discussion of the theoretical structure on which this history is built, see J. B. Buhs, "The Fire Ant Wars: Nature and Science in the Pesticide Controversies of the Late Twentieth Century," *Isis* 93 (2002): 377–400.

6. Quotation from C. J. Bosso, *Pesticides and Politics: The Life Cycle of a Public Issue* (Pittsburgh: University of Pittsburgh Press, 1987), 87.

the entomological practice of exterminating insects with chemical insecticides, and also establishing the power of their bureaucracy.

The South's university entomologists, hunters, wildlife biologists, and early environmentalists articulated a different understanding of the ant and the chemicals used to kill it. The region's entomologists studied the ant in places where its irruption had flagged, and so they saw an insect that caused only minor problems. The ant, they concluded, was not so bad. Some even suggested it might be more beneficial than harmful. This research became a weapon in the battle among entomologists, proof that entomologists needed to study insects, not just the chemicals employed to eradicate them. Others also used the work of the South's entomologists in a spirited critique of the USDA. Many hunters and wildlife biologists contrasted the ambiguity surrounding the ant with what they considered the certainty that insecticides destroyed game animals. They argued that the USDA was a rogue bureaucracy, undermining the public good to fill the coffers of its allies in the chemical industry and extend its own power. Later, environmentalists argued that insofar as the ant became part of the natural balance, it hardly posed any problems at all, but that the insecticides used to kill the bug would ultimately harm humans, passing from the environment through animals and into human bodies, where they would cause disease and death.

Understanding these views as shaped by society does not discredit them. None were manufactured out of whole cloth, but even the best of them was limited, as all ideas about nature are. The ant proved more complicated and more surprising than anyone knew: aspects of its natural history were, literally, unimaginable. Thus, even as environmentalists argued that the insect was entering the natural balance, the insect evolved, and a new form of *Solenopsis invicta* came to dominate the South. It built as many as four hundred mounds on a single acre. It decimated biodiversity, killing native ants and other arthropods. It even threatened bird species that environmentalists were trying to save.

The ideas about *Solenopsis invicta* and about the insecticides, although never perfect representations of nature, became guides for action. In chapter 4, I track how the different groups transformed their ideas into public policy. Should the ant be eradicated? Or should the spraying stop? As the fire ant wars developed, two sides coalesced, and they took turns seeing their ideas made real. The USDA tried to eradicate the ant for over two decades (although the intensity of the attempt fluctuated). For much of that time, environmentalists tried to outlaw the chemicals that the PPC employed.

By 1983, the EPA had banned all the insecticides used in the eradication campaigns. But, just as ideas about nature never fully captured the essence of nature, neither did these policies ever fully embody the intentions of the various groups. There were compromises and concessions.

In chapter 5, I look at the consequences of these two different policies. What did the eradication effort do? What repercussions followed from banning the insecticides used to kill the ant? The USDA's efforts, I argue, failed to check the invasion of the insect—and may have even helped it to spread. The insecticides acted like the rivers of the South American interior or the bulldozer revolution, creating disruption that benefited *Solenopsis invicta*. The EPA's efforts were more successful—dangerous chemicals were removed from the market—but also came alloyed with unintended consequences. Environmentalists hoped that by limiting human interaction with nature, the natural balance could re-assert itself. But humans had already altered nature. Americans had shaped the southern landscape and what they had built favored the ant. The insect continued to spread, proving more pestilent than the USDA's critics had thought. The ant helped to fuel a backlash against the environmental movement and the EPA as people confronted by the nettlesome insect complained that the environmental bureaucracy cared more for the ant than for them. While we humans shape nature, we are not in full control of what we create. The friction between nature, our limited ideas about nature, and how we transform those ideas into reality creates situations that no one can predict.

The lesson, I think, is that nature is neither something that we can manipulate without consequence, as the USDA's failure shows, nor something from which we can remove ourselves, as the EPA's failure shows. We need to recognize that we change the world, but that the consequences of those changes are, and always will be, unpredictable to some extent. It is humanity's blessing and burden: to have the power to change the world and the responsibility for repercussions that we cannot fully anticipate.

I hope, then, that this book contributes to the understanding of the ways humans and nature have interacted in the past. I also hope that it offers some insight, however limited, about what we might expect in the future. The anthropologist Elizabeth Lawrence writes, "It is urgent, if we are ever to establish harmonious relationships with other forms of life on the planet, to understand how the beliefs of a particular culture, in combination with observations about the natural history of an animal, result in a symbolic entity that elicits emotional and intellectual responses that influence society's treatment of that animal and may ultimately determine the fate of the species."[7] I would replace the word "harmonious" with "accommodating," since accommodation connotes the inevitable friction that I see between nature, ideas about nature, and ways of transforming nature. But otherwise this history of the fire ant wars is an attempt to do what Lawrence suggests: to investigate how human history and the natural history of an animal have intertwined, and the consequences of that relationship for both the animal and for us.

7. E. Lawrence, *Hunting the Wren: Transformation of Bird to Symbol* (Knoxville: University of Tennessee Press, 1997), xii.

Chapter One: From South America to the American South, 1900–1950

Out of the night that covers me,
  Black as the Pit from pole to pole,
I thank whatever gods may be
  For my unconquerable soul.

William Ernest Henley, "Invictus"

The red imported fire ant, noted journalist Charles Haddad in 1990, was not only "a part of the South's ecosystem but" was "embedded in folklore as well."[1] And not just folklore: the ant was a part of southern culture. By the end of the twentieth century the insect occupied about 300 million acres in North America, from Florida to California and as far north as Virginia, an entrenched and ubiquitous denizen of the Sun Belt. Musicians wrote songs about the insect, sports teams used it as a mascot, and the town of Marshal, Texas dedicated an annual festival to it. The ant figured in advertisements, political speeches, and novels. "The fire ant is a swarming, biting nasty little critter whose legend is to flatlands what bigfoot is to the mountains," the *Atlanta Journal-Constitution* proclaimed in 1986.[2] But this was not always the case. The ant evolved in South America and only came to the American South during the first half of the twentieth century. How did the insect reach North America? And how did it spread and come to dominate the South?

The fire ant's history is a tale of nature, humans, and their interactions. Evolving on a South American floodplain, the fire ant became an adept tramp, able to move easily and exploit open, disturbed habitats. Then, in the late nineteenth and early twentieth centuries, shipping routes tethered its homeland to world markets. Accidentally loaded aboard boats with South American cargo, *Solenopsis invicta* found itself in Mobile, Alabama, subsequently spreading from the port city across American South. The fire ant's spread, I argue, resulted from the interaction between the insect's biology and human activities. When the imported fire ant reached the United States, American citizens were busily recasting the South's economy and ecology, creating new

1. C. Haddad, "Fire Ants Defy Eradication, Burrow into Southern Lore," *Atlanta Journal-Constitution,* 11 November 1990, B1.
2. B. Harrell, "Mounds of Trouble from the Fire Ant: When Southerners Talk about the Red Scourge, They Don't Mean Russia," *Atlanta Journal-Constitution,* 6 July 1986, M10.

avenues of transport as well as new open and disturbed areas. The ant traveled along these new corridors and exploited the new habitats. Natural history and human history intertwined to produce the imported fire ant's explosion.

Natural History of the Imported Fire Ants

The imported fire ant belongs to the genus *Solenopsis,* a group of some two hundred species of ants; only a handful of these are fire ants, about twenty, the uncertainty reflecting the limited knowledge we have of some species.[3] The genus originated in tropical South America around 65 million years ago, as the dinosaurs made their last stand and the Andean orogeny proceeded along the continent's western spine. According to the geological record, *Solenopsis* underwent an adaptive radiation during the Miocene, before the first *Homo sapiens* walked across the African savannah. A variety of forms evolved. Some *Solenopsis* eat seeds; some parasitize other ants. Some live along the littoral, some in treetops, some in subterranean tunnels.[4] Those ants that came to be known as "fire ants" developed a potent sting that gave the group its common name. The humorist Dave Barry remembered that the first time he was stung by a fire ant he "leaped up and danced wildly around, brushing uselessly at my hand, which felt as though I had stuck it in a toaster oven set on 'pizza.'"[5] But while the insects developed different lifestyles, changes in their appearance did not always keep pace. As a group, the ants are rather nondescript, and many species look alike, even to experts.[6] The ant biologist William Steel Creighton wrote that when studying the insects' taxonomy, "[o]ne has the unpleasant feeling that he is entering a battle-field strewn with unexploded missiles and that there is a strong probability that one of these taxonomic duds may, through tampering, bring the investigator to grief."[7]

This confusion clouds the early history of the imported fire ant invasion. For many years, the name "imported fire ant" covered two species, *Solenopsis richteri* and *Solenopsis invicta,* both of which jour-

---

3. J. C. Trager, "A Revision of the Fire Ants, *Solenopsis geminata* Group (Hymenoptera: Formicidae: Myrmicinae)," *Journal of the New York Entomological Society* 99 (1991): 146.

4. On the evolutionary history of *Solenopsis,* see E. O. Wilson, "The Defining Traits of Fire Ants and Leaf-Cutting Ants," in *Fire Ants and Leaf-Cutting Ants: Biology and Management,* ed. C. S. Lofgren and R. K. Vander Meer (Boulder: Westview, 1986), 2–4.

5. D. Barry, "Forget about the Axis of Evil—Let's Get Rid of the Fire Ant," *Sacramento Bee,* 4 May 2003, L2.

6. E. O. Wilson, personal communication.

7. W. S. Creighton, "The New World Species of the Genus *Solenopsis* (Hymenoptera: Formicidae)," *Proceedings of the American Academy of Arts and Sciences* 66 (1930): 69.

neyed from South America to the American South during the first half of the twentieth century. *Solenopsis richteri* is a dark brown or black species, with a yellow stripe across its gaster. *Solenopsis invicta* lacks the stripe of its congener and is red, although sometimes black or brown, too. The two species inhabit different parts of the world's largest wetland, an expanse of marshy land that follows the Río Paraguay through Brazil, Paraguay, and northern Argentina to its confluence with the Río Paraña—where the drainage is known as the Río de la Plata—and, ultimately, to the Atlantic Ocean. The black imported fire ant lives at the southern edge of this riverine system, often in seasonally waterlogged grasslands; it also lives along roadsides in Buenos Aires and amid the grasslands of the drier Argentine pampas. The range of the red imported fire ant runs north from Buenos Aires, following the Paraguay and Paraña into Brazil. The ant also survives on the harsh plains of the Chaco.[8] *Solenopsis invicta* is the insect now known as the "imported fire ant," the one that has caused most of the problems and has provoked both the South's loathing and pride.

The rivers and their tributaries set the pulse for the ants' South American home. During the dry season, from April to October, rivers run weakly and grasses clog the riverbeds, growing thick in the rich silt deposited by the rivers in years past. When the rains come, and the waters tumble from the Brazilian plateau to the north and the Andes to the west, the tangle of flora forces the rivers to cut new channels. These fill with silt and eventually overflow, flooding the countryside for months. The rivers' vagrancy has created a wealth of habitats and the area supports a rich array of plant life.[9] In 1927 one geographer noted, "The most striking feature in the natural vegetation is its lack of uniformity."[10] Palms stand in swamps; tall grasses dominate savannah-like stretches, while shorter grasses survive on areas that have been more recently covered with water. In parts of the Chaco, quebracho trees grow into forests.

Solenopsis invicta evolved to take advantage of the open and disturbed habitats left in the wake of floods. The entomologist Walter Tschinkel compared the insect to a weed. The ant, Tschinkel said,

8. S. W. Taber, *Fire Ants* (College Station: Texas A&M Press, 2000), 25–28, 58–60.

9. On the ecology of the region, see A. A. Bonettos and I. R. Wais, "Southern South American Streams and Rivers," in *River and Stream Ecosystems*, ed. C. E. Cushing, K. W. Cummins, and G. W. Minshall, vol. 22 of *Ecosystems of the World*, ed. D. W. Goodall (Amsterdam: Elsevier, 1995), 257–93.

10. E. W. Shanahan, *South America: An Economic and Regional Geography with an Historical Chapter* (London: Methuen, 1927), 90.

breeds rapidly, spreads easily, and competes fiercely.[11] A single red queen gives birth to a colony of 230,000 in about three years and produces about 4,000 to 6,000 reproductively viable offspring (called "alates") annually.[12] The alates are produced throughout the year, ready to spread from the nest whenever conditions are right.[13] On clear, calm days after a rain storm, the red alates climb atop the mound then fly away, mating in the air; the males drop dead, but the females fly on, some as far as twelve miles, although most do not cover such a distance.[14] Watching a mating frenzy in the United States during the spring of 1951, an entomologist for the U.S. Department of Agriculture (USDA) was shocked at the ants' "tremendous biotic potential."[15] The ant also uses floods to spread: when waters overwhelm a nest, the colony forms a mass, floats to a dry spot and rebuilds.[16]

Whether alighting after a nuptial flight or borne aloft by floodwaters, *Solenopsis invicta* settles in any number of environments. It succeeds best, however, in open or disturbed habitats, the kind of places created by the floods and the rivers' retreat. Tschinkel notes, "The fire ant is clearly and dramatically associated with ecologically disturbed habitats. . . . On the other hand, the fire ant is absent or rare in late succession or climax communities such as mature deciduous or pine forest. When it is found in these communities, it is usually associated with local disturbances such as seasonal flooding and roads."[17] In these

11. W. R. Tschinkel, "History and Biology of Fire Ants," in *Proceedings of the Symposium on the Imported Fire Ant,* ed. S. L. Battenfield (Washington, D.C.: Government Printing Office, 1982), 21. See also W. R. Tschinkel, "The Ecological Nature of the Fire Ant: Some Aspects of Colony Function and Some Unanswered Questions," in *Fire Ants and Leaf-Cutting Ants: Biology and Management,* ed. C. S. Lofgren and R. K. Vander Meer (Boulder: Westview, 1986), 72.

12. C. S. Lofgren, W. A. Banks, and B. M. Glancey, "Biology and Control of Imported Fire Ants," *Annual Review of Entomology* 20 (1975): 8; S. B. Vinson, "Invasion of the Red Imported Fire Ant (Hymenoptera: Formicidae): Spread, Biology, and Impact," *Bulletin of the Entomological Society of America* 43 (1997): 27.

13. D. P. Wojick, "Observations on the Biology and Ecology of Fire Ants in Brazil," in *Fire Ants and Leaf-Cutting Ants: Biology and Management,* ed. C. S. Lofgren and R. K. Vander Meer (Boulder: Westview, 1986), 92; W. R. Tschinkel, "Sociometry and Sociogenesis of Colonies of the Fire Ant *Solenopsis invicta* during One Annual Cycle," *Ecological Monographs* 64 (1993): 441–42.

14. G. P. Markin, J. H. Dillier, S. O. Hill, M. S. Blum, and H. P. Hermann, "Nuptial Flights and Flight Ranges in the Imported Fire Ant, *Solenopsis saevissima richteri,*" *Journal of the Georgia Entomological Society* 6 (1971): 145–50.

15. L. F. Lewis, "Investigations in the Biology of the Imported Fire Ant *Solenopsis saevissima* var. *richteri* Forel," fire ant file, SG 9991, Alabama Department of Conservation, director's administrative files, Alabama Department of Archives and History, Montgomery, Ala. (hereafter ALDoC papers).

16. W. L. Morrill, "Dispersal of Red Imported Fire Ants by Water," *Florida Entomologist* 57 (1974): 39–42.

17. W. R. Tschinkel, "The Ecological Nature of the Fire Ant: Some Aspects of Colony Function and Some Unanswered Questions," in *Fire Ants and Leaf-Cutting Ants: Biology and Management,* ed. C. S. Lofgren and R. K. Vander Meer (Boulder: Westview, 1986), 72–73.

exposed areas, the insect protects itself from predators by building large dirt mounds. Its sting is also a useful adaptation for pioneering disturbed environments. The venom produced by fire ants, unlike poisons produced by most other animals, contains alkaloids, chemicals more commonly found in plants.[18] In mammals the venom kills cells; it is also a bactericide, fungicide, herbicide, and insecticide.[19] Many fire ant alates still succumb to predators, but since colonies release so many alates, even if only a small percentage survive the number of colonies in an area grows.[20] The ant's catholic diet also helps it to survive in disturbed areas where food sources are not reliable. *Solenopsis invicta* eats anything from plants to insects to vertebrates, although it seems to prefer insects and oily foods.[21] (In Mato Grosso, Brazil, locals call the red ant by the Portuguese name *toicinhera*, derived from the word *toicinho*—pork fat. There, *Solenopsis invicta* gorges on the oily nuts of the Baçu palm.)[22]

The red ant also seems to have another advantage for colonizing new and disturbed areas: its larvae. Food brought to the nest by workers passes to the larvae, which digest some things that other ants in the nest either cannot or will not. The metabolic by-products of the larvae's digestion are then passed to the queen. If these waste products contain too-low a level of important nutrients, the queen's reproductive output

18. J. G. MacConnell and M. S. Blum, "Alkaloid from Fire Ant Venom: Identification and Synthesis," *Science* 168 (1970): 840–41.

19. M. S. Blum, J. R. Walker, P. S. Callahan, and A. F. Novak, "Chemical, Insecticidal, and Antibiotic Properties of Fire Ant Venom," *Science* 128 (1958): 306–7; J. T. Sinski, G. A. Adrouny, V. J. Derbes, and R. C. Jung, "Further Characterization of Hemolytic Component of Fire Ant Venom, Mycological Aspects," *Proceedings of the Society for Experimental Biology and Medicine* 102 (1959): 659–60; G. A. Adrouny, V. J. Derbes, and R. C. Jung, "Isolation of a Hemolytic Component of Fire Ant Venom," *Science* 130 (1959): 449; P. Escoubas and M. S. Blum, "The Biological Activities of Ant-Derived Alkaloids," in *Applied Myrmecology: A World Perspective*, ed. R. K. Vander Meer, K. Jaffe, and A. Cedeno (Boulder: Westview, 1990), 482–89.

20. W. H. Whitcomb, A. P. Bhatkar, and J. C. Nickerson, "Predators of *Solenopsis invicta* Queens Prior to Successful Colony Establishment," *Environmental Entomology* 2 (1973): 1101–3; B. J. Nichols and R. W. Sites, "Ant Predators of Founder Queens of *Solenopsis invicta* (Hymenoptera: Formicidae) in Central Texas," *Environmental Entomology* 20 (1991): 1024–29.

21. On fire ant food preferences in Brazil, see D. P. Wojick, "Observations on the Biology and Ecology of Fire Ants in Brazil," in *Fire Ants and Leaf-Cutting Ants: Biology and Management*, ed. C. S. Lofgren and R. K. Vander Meer (Boulder: Westview, 1986), 93; on their preferences in the United States, see J. T. Vogt, R. A. Grantham, E. Corbett, S. A. Rice, and R. E. Wright, "Dietary Habits of *Solenopsis invicta* (Hymenoptera: Formicidae) in Four Oklahoma Habitats," *Environmental Entomology* 31 (2002): 47–53. On their attraction to fatty foods, see S. B. Vinson, J. L. Thompson, and H. B. Green, "Phagostimulants for the Imported Fire Ant, *Solenopsis saevissima* var. *richteri*," *Journal of Insect Physiology* 13 (1967): 1729–36.

22. F. E. Lennartz, "Modes of Dispersal of *Solenopsis invicta* from Brazil into the Continental United States—A Study in Spatial Diffusion" (Master's thesis, University of Florida, 1973), 217.

is curtailed: there is no reason to produce young that cannot be fed. Alternatively, some by-products increase the rate at which queens lay eggs. This system allows the ant to calibrate its biology with local conditions, retrenching when resources are scarce and expanding to take advantage of favorable conditions, spreading quickly and easily, like a weed.[23]

While life along a flood plain sculpted the red imported fire ant into the insect equivalent of a weed, *Solenopsis invicta* is not the only opportunistic member of *Solenopsis*. *Solenopsis geminata,* the tropical fire ant, has spread to every continent except Antarctica and Europe, traveling aboard ships and exploiting the disturbed, simplified, and open environments created by humans. It is widely distributed across the southern tier of the United States.[24] Beginning in the sixteenth century, an ant believed to be *Solenopsis geminata* plagued the islands of the Caribbean episodically; the insect caused so many problems that Spanish colonists considered abandoning Hispaniola and officials in Grenada offered a £20,000 award for anyone who invented a way to exterminate it.[25] In the nineteenth century the English naturalist Henry Walter Bates watched another species of fire ant terrorize the village of Aveyros, on the Tapajos River in the Amazon basin. Years before Bates arrived, martial strife had devastated the area and many had fled their homes.[26] They returned, were driven out by the ant, then returned again, only to see the insect still undermining their village.[27] "The ground is perforated with entrances to [the ant's] subterranean galleries, and a little sandy dome occurs here and there," Bates wrote. "The houses are overrun with them; they dispute every fragment of food with the inhabitants."[28] The ant, he said, was "a greater plague than all the other [insects] put together" and because of them "all eatables are obliged to be suspended in baskets from the rafters, and the cords well soaked with copaüba, which is the only means known of

23. W. R. Tschinkel, "Social Control of Egg-Laying Rate in Queens of the Fire Ant, *Solenopsis invicta,*" *Physiological Entomology* 13 (1988): 327–50; W. R. Tschinkel, "Stimulation of Fire Ant Queen Fecundity by a Highly Specific Brood Stage" *Annals of the Entomological Society of America* 88 (1995): 876–82; D. L. Cassill and W. R. Tschinkel, "Effects of Colony-Level Attributes on Larval Feeding in the Fire Ant, *Solenopsis invicta,*" *Insectes Sociaux* 46 (1999): 261–66.

24. S. W. Taber, *Fire Ants* (College Station: Texas A&M Press, 2000), 66–70.

25. F. Cowan, *Curious Facts in the History of Insects* (Philadelphia: Lippincott, 1865), 166–69; R. A. F. Réaumur, *The Natural History of Ants,* ed. and trans. W. M. Wheeler (New York: Knopf, 1926), 232–39; E. O. Wilson, *The Insect Societies* (Cambridge: Harvard University Press, 1971), 451.

26. H. W. Bates, *The Naturalist on the River Amazons* (London: John Murray, [1863] 1892), 252.

27. Ibid., 251.

28. Ibid., 252.

preventing them from climbing. They seem to attack persons out of sheer malice."[29]

For the villagers, the ant symbolized the ruins on which their city was rebuilt, the past come to life. The locals, Bates wrote, believed "that the hosts sprang up from the blood of the slaughtered."[30] The connection between the war and the ant must have seemed especially strong since the ant rarely moved beyond the borders of the village. Aveyros occupied a small clearing, hemmed in a by a thick, dark tropical forest. Preferring the open disturbed habitat to the tightly interwoven ecological community below the jungle canopy, the ant was found only in the village and along the sandy shores of the river, where the war had raged and the dead had lain upon the ground.[31]

Bates recorded his observation in *The Naturalist on the River Amazons,* published in 1863. The book became a classic of adventure and natural science, read throughout the Anglo-American world. This ant, Bates implied, was the antithesis of civilization, growing up in its absence, challenging its return. Bates's judgment was reflected in the insect's name. The traveler captured some specimens of the offending ant and sent them to the British Museum in London, where they were pinned, placed in a collection, and studied. The ant was dubbed *Solenopsis saevissima:* savage fire ant.[32]

From South America to the American South

As *Solenopsis saevissima* terrorized Aveyros and *Solenopsis geminata* tramped around the globe, *Solenopsis richteri* and *Solenopsis invicta* remained confined to South America. How did they reach North America? Their passages went unrecorded, but the history can be pieced together. The reconstruction shows that while the ants' biology mattered, it only partly explains their later success: the explosion of fire ants across the American South depended on human commerce and, it seems, the accidental introduction of another insect into the area as well.

Much of the area occupied by *Solenopsis richteri* and *Solenopsis invicta* is uninviting to humans and remained lightly occupied into the nineteenth

29. Ibid., 251–52.

30. Ibid., 252.

31. Ibid., 251–52. On Bates, see G. Woodcock, *Henry Walter Bates: Naturalist of the Amazons* (London: Farber and Farber, 1969).

32. On the taxonomic history of the insect, see W. M. Wheeler, "Some Additions to the North American Ant-Fauna," *Bulletin of the American Museum of Natural History* 34 (1915): 389–421; W. M. Wheeler, "Notes on the Brazilian Fire Ant *Solenopsis saevissima* F. Smith," *Psyche* 23 (1916): 142–43; W. S. Creighton, "The New World Species of the Genus *Solenopsis* (Hymenoptera: Formicidae)," *Proceedings of the American Academy of Arts and Sciences* 66 (1930): 69–73, 80, 87–89.

century: during the dry season when the floods recede and the rain stops, rivers dry to a trickle and water is scarce, found mostly in brackish pools. Daytime temperatures climb over 100°F and at night mosquitoes materialize out of the grass. "Few areas have a climate so extreme and unforgiving that visitors regularly use the word 'hell' when describing it," commented science writer Erich Hoyt in 1996, but the South American interior deserved the name.[33] Increasing familiarity with the area during the Paraguayan War (1864–70), though, drew people to the area, and an influx of European immigrants also helped to extend agriculture and cattle ranching into the hinterlands of Brazil and Argentina.[34] Railroads stretched from Buenos Aires, Rosario, and São Paulo into the swampy continental interior: between 1895 and 1914, for example, Argentina laid over twelve thousand miles of tracks and cargo weight increased from 5 to 40 million tons.[35] Quebracho trees provided fuels for trains and tannins to cure cowhides into leather.[36] Other construction projects made the harsh climate habitable. Irrigation canals, swamp draining, and dams controlled the mosquito population, assured the permanence of pasture, and provided a steady supply of fresh water. Farmers sowed fields with tobacco and cotton.[37]

Over time, Argentina and Brazil became two of the world's leading beef-producing countries. In the late nineteenth century, Argentineans had shipped live cattle to Europe and America. After a long voyage, however, the animals arrived sickened and weak; few sold. An outbreak of hoof-and-mouth disease among the Argentinean herds around the turn of the century forced changes in the industry. Argentineans started sending the beef frozen or in cans, making shipping less risky. They also imported blooded cattle from England and France to better meet the tastes of their consumers. As a result of these changes, the industry boomed. By 1935, Argentina provided over half the world's

33. E. Hoyt, *The Earth Dwellers: Adventures in the Land of Ants* (New York: Touchstone Books, 1996), 16.

34. On the region's hydrology, see A. A. Bonettos and I. R. Wais, "Southern South American Streams and Rivers," in *River and Stream Ecosystems*, ed. C. E. Cushing, K. W. Cummins, and G. W. Minshall, vol. 22 of *Ecosystems of the World*, ed. D. W. Goodall (Amsterdam: Elsevier, 1995), 257–93. On the role of the Paraguayan War, see R. W. Wilcox, "'The Law of Least Effort': Cattle Ranching and the Environment in the Savanna of Mato Grosso, Brazil, 1900–1980," *Environmental History* 4 (1999): 343. On the peopling of the region, see R. C. Eidt, *Pioneer Settlement in Northeast Argentina* (Madison: University of Wisconsin Press, 1971).

35. D. Rock, *Argentina 1516–1982: From Spanish Colonization to the Falklands War* (Berkeley: University of California Press, 1985), 169.

36. E. W. Shanahan, *South America: An Economic and Regional Geography with an Historical Chapter* (London: Methuen, 1927), 194–200.

37. A. Morris, *South America* (New Jersey: Barnes and Noble Books, 1987), 143.

beef.[38] Meanwhile, World War I prodded the growth of Brazil's cattle industry. Frozen and chilled beef exports grew from 1.5 metric tons in 1914 to 65,000 metric tons in 1917.[39]

The railroads, farms, and ranches imposed a new ecological order on the region, one that operated not according to the rhythms of the wet and dry seasons, nor in step with the *pas de deux* danced by the region's hydrology and botany, but according to human imperatives. New species thrived—cattle, grasses imported to feed the animals—while the populations of some plants and animals long established in the region struggled.[40] But while humans had created the system, they were not always in control: humans change nature, but nature always remains more complicated than humans know, and the manifold consequences of our actions are impossible to fully predict. Thus, the ships that visited the port cities of eastern South America carried not only cattle and agricultural goods to the world market, but *Solenopsis richteri* and *Solenopsis invicta* as well, travelers that no one saw or imagined.

Not expecting to see *Solenopsis richteri* or *Solenopsis invicta*, no one recorded their passage from South America to North America. Nonetheless, a likely scenario can be reconstructed from what is known. The black imported fire ant, closer to shipping routes than its congener, seems to have made its way north first, some time around 1918 when Mobile nursery owner and amateur entomologist Henry Peter Löding found it girdling his satsuma trees.[41] Years later, some Alabamans claimed to have seen the ant around the dock as early as the turn of the century, but they lacked entomological training and were relying on memories almost five decades old, so the recollections cannot be taken as more than suggestive.[42] And even if correct, the reports

38. On changes in the Argentinean cattle industry, see Y. F. Rennie, *The Argentine Republic* (Westport, Conn.: Greenwood Press, 1945), 142–50. Argentina's 1935 ranking among beef-producing nations is from P. P. Egoroff, *Argentina's Agricultural Exports during World War II* (Palo Alto: Food Research Institute, Stanford University, 1945), 1.

39. Statistics and history of the Brazilian cattle industry are from R. W. Wilcox, "'The Law of Least Effort': Cattle Ranching and the Environment in the Savanna of Mato Grosso, Brazil, 1900–1980," *Environmental History* 4 (1999): 343–44.

40. On the region's ecological changes, see V. Banks, *The Pantanal: Brazil's Forgotten Wilderness* (San Francisco: Sierra Club Books, 1991).

41. On the discovery of the ant, see H. P. Löding, "An Ant," *U. S. Department of Agriculture Insect Pest Survey Bulletin* 9 (1929): 241; W. S. Creighton, "The New World Species of the Genus *Solenopsis* (Hymenoptera: Formicidae)," *Proceedings of the American Academy of Arts and Sciences* 66 (1930): 88–89. On Löding, see "Dr. H. P. Löding of Mobile, Ala., 1869–1942," *Journal of the New York Entomological Society* 37 (1942): 50–51.

42. On these early reports, see L. F. Lewis, "Investigations in the Biology of the Imported Fire Ant," 1951, fire ant file, SG 17019, ALDoC papers; F. E. Lennartz, "Modes of Dispersal of *Solenopsis invicta* from Brazil into the Continental United States—A Study in Spatial Diffusion" (Master's thesis, University of Florida, 1973), 211–15.

add little to an understanding of the black imported fire ant's North American history. Since an introduction date around 1918 accords with the history of South America's export industry, it is safer to assume that the ant reached North America sometime toward the end of World War I.

According to chemical analyses, the ant came from an area one hundred kilometers north of Buenos Aires, near the Uruguay border.[43] No one is sure which ship, or ships, carried the insect. The bigger question, though, is why the ant established itself in Mobile. The Alabama city was a major port, one of the chief links between eastern South America and the American South, but New Orleans received more cargo from the fire ant's homeland than Mobile.[44] So why did the black ant establish itself in Alabama and not Louisiana?

In 1973, the ant taxonomist William Buren and some colleagues hypothesized that the solution to this conundrum lay in the accidental introduction of another insect—the Argentine ant (*Linepithema humile*, formerly known as *Iridomyrmex humilis*).[45] Like fire ants, the Argentine ant is a tramp that has spread over the globe displacing native ants; but, unlike *Solenopsis richteri* or *Solenopsis invicta*, the Argentine ant prefers sugary foods and is frequently found in or near homes.[46] Apparently, the Argentine ant reached New Orleans around the turn of the century, carried by a ship loaded with coffee. The ant quickly spread through the city and the USDA initiated a program to control the insect.[47] Buren suggested that the Argentine ant prevented the black imported fire ant from establishing itself in New Orleans: *Solenopsis richteri* is a weedy species, but it could not find a place in an ant community that the Argentine ant dominated.

43. R. K. Vander Meer and C. S. Lofgren, "Chemotaxonomy Applied to Fire Ant Systematics in the United States and South America," in *Applied Myrmecology: A World Perspective*, ed. R. K. Vander Meer, K. Jaffe, and A. Cedeno (Boulder: Westview, 1990), 81.

44. W. F. Buren, G. E. Allen, W. H. Whitcomb, F. E. Lennartz, and R. N. Williams, "Zoogeography of the Imported Fire Ants," *Journal of the New York Entomological Society* 82 (1974): 115.

45. Ibid., 113–24.

46. For a discussion of the Argentine ant's peregrinations, see W. M. Wheeler, "On Certain Tropical Ants in the United States," *Entomological News* 17 (1906): 23–26. For a discussion of the Argentine ant's habits, see M. R. Smith, *House-Infesting Ants of the Eastern United States: Their Recognition, Biology, and Economic Importance*, Technical Bulletin 1326 (Washington, D.C.: Agricultural Research Service, USDA, 1965), 54.

47. On the Argentine ant's invasion, see W. Newell, "Notes on the Habits of the Argentine or 'New Orleans' Ant, *Iridomyrmex humilis* Mayr," *Journal of Economic Entomology* 1 (1908): 20–34; W. Newell, "The Life History of the Argentine Ant," *Journal of Economic Entomology* 2 (1909): 174–92; W. Newell and T. C. Barber, "The Argentine Ant," *USDA Bureau of Entomology Bulletin* 122 (1913): 1–98.

The situation in Mobile, though, was different. As early as 1915, the Argentine ant crawled over Alabama soils in sufficient numbers to persuade the agricultural experiment station to spread baits of meat, honey, or sugar mixed with arsenic, cyanide, or Paris green around Montgomery and Birmingham.[48] No complaints emanated from Mobile, though. Presumably only a relatively small number of Argentine ants occupied the city, if any at all. Faced with less competition, the fire ant became entrenched.

The insect's position, however, remained precarious. In the 1920s, the Argentine ant population exploded in Mobile, and the composition of the ant community changed, although exactly how remains unclear.[49] Löding reportedly said that *Linepithema humile* forced the black imported fire ant north of the city; but another observer, William Steel Creighton, disagreed.[50] A biology student at Roanoke College who had become interested in ants after reading the works of William Morton Wheeler, a Harvard professor and the world's leading ant biologist, Creighton spent summers in Mobile, where his parents lived.[51] The Argentine ant, he noticed, did displace *Solenopsis richteri* (as well as native ants), but did not push it beyond Mobile's borders. He found the black imported fire ant haunting a few blocks near the Government Street Loop, a neighborhood along the Mobile River where the trolleys turned around.[52]

The red imported fire ant arrived some time after 1933, when Creighton—who never found the red ant in Mobile—stopped collecting in the area, and before 1945, when Buren found it a few miles from the port (although he did not recognize the insect at the time). It may have been during the early years of World War II, since closed European ports meant that more cargo from South America reached the United

48. For the Argentine ant's early history in Alabama, see E. A. Vaughn to Davis, 9 November 1915; E. A. Vaughn to Davis, 15 April 1916; B. M. Parks to Health Department, 25 April 1916; Berry to E. A. Vaughn, 3 May 1916; H. F. Malone to J. D. Dowling, 16 March 1923; J. C. Rivers to W. E. Hinds, 23 March 1923, all in file 17, box 1, Warren E. Hinds papers, collection 615, Auburn University Archives.

49. The explosion of the Argentine ant population in Mobile is documented in W. S. Creighton to M. S. Blum, 22 April 1968, William Steel Creighton papers, an unprocessed box of material incorporated in the uncatalogued E. O. Wilson papers, Library of Congress, Washington, D.C. (hereafter Creighton papers).

50. On Löding's opinion, see E. O. Wilson, "Variation and Adaptation in the Imported Fire Ant," *Evolution* 5 (1951): 68.

51. On Creighton, see J. B. Buhs, "Building on Bedrock: William Steel Creighton and the Reformation of Ant Systematics, 1925–1970," *Journal of the History of Biology* 33 (2000): 27–70.

52. W. S. Creighton to M. S. Blum, 22 April 1968, Creighton papers.

States.[53] Chemical studies indicate that the ant came from along the Río Paraguay drainage and genetic tests suggest that between five and fifteen immigrant queens gave rise to the billions that would spread across the American South in coming decades.[54] But no one is sure how the queens came to North America.[55] Perhaps they arrived in fruit ships.[56] Perhaps they traveled in ships loaded with cattle or beef.[57] Perhaps they were mixed with ballast: at the time ships from South America that carried relatively light agricultural goods were weighed down with soil; when the boats took on heavy machinery in Mobile for transport back to South America, the ballast was dumped. *Solenopsis invicta* may have been dumped along with it.[58]

The red imported fire ant almost certainly reached New Orleans as well as Mobile, but in Louisiana, *Solenopsis invicta,* as *Solenopsis richteri* before it, met the Argentine ant. *Solenopsis invicta* is a fierce competitor, but it was outnumbered and probably succumbed to the entrenched Argentine ant. In Mobile, though, the black imported fire ant had withstood the onslaught of Argentine ants and, in Buren's words, may have "helped to alleviate some of the competition and predation from the native ants and from *I. humilis,* while at the same time keeping an ecological niche partly open."[59] The red imported fire ant exploited this niche and established itself in Mobile. Battles between the fire ants and the Argentine ants continued, but by the 1940s both *Solenopsis richteri* and *Solenopsis invicta* had found a place in North America.[60]

53. Buren and Creighton discuss their collecting histories in W. F. Buren to W. S. Creighton, 15 January 1969; W. F. Buren to W. S. Creighton, 29 January 1969; W. S. Creighton to F. E. Lennartz, 31 March 1973, all in Creighton papers.

54. R. K. Vander Meer and C. S. Lofgren, "Chemotaxonomy Applied to Fire Ant Systematics in the United States and South America," in *Applied Myrmecology: A World Perspective,* ed. R. K. Vander Meer, K. Jaffe, and A. Cedeno (Boulder: Westview, 1990), 81; S. W. Taber, *Fire Ants* (College Station: Texas A&M Press, 2000), 16.

55. F. E. Lennartz, "Modes of Dispersal of *Solenopsis invicta* from Brazil into the Continental United States—A Study in Spatial Diffusion" (Master's thesis, University of Florida, 1973), reviewed a host of possible routes from Brazil but could not isolate a likely source.

56. G. H. Blake Jr., "Imported Fire Ant—on the March in Alabama," *Highlights of Agricultural Research* 3 (1956): 1.

57. J. C. Trager, "A Revision of the Fire Ants, *Solenopsis geminata* Group (Hymenoptera: Formicidae: Myrmicinae)," *Journal of the New York Entomological Society* 99 (1991): 176.

58. S. B. Vinson, "Invasion of the Red Imported Fire Ant (Hymenoptera: Formicidae): Spread, Biology, and Impact," *Bulletin of the Entomological Society of America* 43 (1997): 25. Creighton suggested the same: W. S. Creighton to M. S. Blum, 14 May 1968, Creighton papers.

59. W. F. Buren, G. E. Allen, W. H. Whitcomb, F. E. Lennartz, and R. N. Williams, "Zoogeography of the Imported Fire Ants," *Journal of the New York Entomological Society* 82 (1974): 116.

60. On the continued battles between the ant species, see L. F. Lewis, "Investigation in the Biology of the Imported Fire Ant, *Solenopsis saevissima* var. *richteri* Forel," fire ant file, SG 9991, ALDoC papers.

Having passed from South America to the American South, the two species of fire ants confronted a new environment that followed a rhythm different than the South American interior. The climate was about the same—Mobile received roughly the same annual rainfall and had the same range of temperatures as the ants' homeland—but, since Alabama and Argentina are on opposite sides of the equator, the seasons were reversed.[61] The ants' biology, though, could handle the inversion of the seasons. Both *Solenopsis richteri* and *Solenopsis invicta* produce alates throughout the year, ready to take to the skies on a warm, calm day after a rain, mate, and propagate the species.[62] During the ants' first northern summer, then—although their biological clocks would have been set for winter—some alates were prepared to climb atop the formicary and set sail. Most likely, fewer ants took part in the mating ritual during the first few years of residence in the United States than during comparable years in the South American interior. Over time, however, their reproductive biology synchronized with the new temporal rhythms, perhaps, in *Solenopsis invicta*'s case, aided by the larvae's sensitive registering of the food available to the colony. The insects seemed robust enough to survive whatever obstacles their new home offered. In the winter of 1928 Creighton was "surprised to find [*Solenopsis richteri*] quite active with brood in the upper most passages at low temperatures."[63]

The Imported Fire Ants and the Bulldozer Revolution
While the two species of imported fire ants had found a place in the American South, they survived on the fringe of the ant community. They spread slowly and attracted only sporadic notice, causing problems unevenly across the South (fig. 2).[64] Farmers in southern Mississippi considered the ants—some mixture of *Solenopsis richteri* and *Solenopsis invicta*—to be among the worst insect pests: their mounds grew so large

61. For climatic comparisons, see F. E. Lennartz, "Modes of Dispersal of *Solenopsis invicta* from Brazil into the Continental United States—A Study in Spatial Diffusion" (Master's thesis, University of Florida, 1973), 124–25.

62. For alate production in the red imported fire ant, see W. R. Tschinkel, "Sociometry and Sociogenesis of Colonies of the Fire Ant *Solenopsis invicta* during One Annual Cycle," *Ecological Monographs* 64 (1993): 441–42. For alate production in the black imported fire ant, see P. R. San Martin, "The Venomous Ants of the Subgenus *Solenopsis*," in *Venomous Animals and Their Venoms*, vol. 3, ed. W. Bücherl and E. E. Buckley (New York: Academic Press, 1971), 101. San Martin's data is for the ant in South America, but presumably it holds true in North America as well.

63. W. S. Creighton to W. M. Wheeler, 2 January 1928, C Regular 1928 file, William Morton Wheeler papers, HUG (FP) 87.10, Harvard University Archives.

64. The ants' march is described as slow in E. O. Wilson and J. H. Eads, "Special Report to the Alabama Department of Conservation," 16 July 1949, Blum papers.

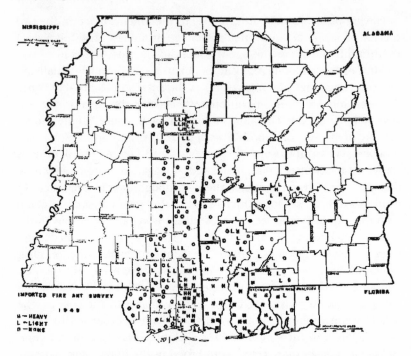

Figure 2. A 1949 USDA survey found the imported fire ant concentrated in southern Alabama, Florida, and Mississippi. The population in northeast Mississippi was near a railroad stop for a line that ran out of Mobile. At the time, no one regularly distinguished between the black and red imported fire ant. (From W. G. Bruce, J. M. Coarsey Jr., M. R. Smith, and G. H. Culpepper, "Survey of the Imported Fire Ant, *Solenopsis saevissima* var. *richteri*, Special Report S-15, July 1949, fire ant file, SG 9998, ALDoC papers.)

and so densely that from a distance they looked like flocks of sheep.[65] And, in 1937, Alabama conducted a program to control fire ants.[66] (Whether the targeted ant was *Solenopsis richteri*, *Solenopsis invicta*, or a native fire ant is impossible to tell.) But other farmers considered the ants a minor nuisance that was easily controlled by frequent cultivation of fields. Only when plows first cut through uncultivated land did the mounds appear, they said. As the land was worked, the insects caused fewer problems.[67] And Creighton never found the insects to be more than a nuisance. Visiting Alabama every year between 1928 and 1933 to collect ants for his dissertation, Creighton kept a stock of

65. M. R. Smith, "Report on Imported Fire Ant Investigations," 26 July 1949, Blum papers.
66. Minutes of the Fire Ant Conference, 20 May 1949, fire ant infestation file, SG 9998, ALDoC papers.
67. M. R. Smith, "Report on Imported Fire Ant Investigations," 26 July 1949, Blum papers.

cyanide on hand to poison colonies of the imported fire ant when his neighbors asked for help; but four decades later, he could not remember dowsing more than two dozen nests.[68]

Then, in the 1950s, the red imported fire ant's population suddenly and dramatically increased while the black ant's seemed to decrease.[69] By 1957, the latecomer ranged over more than 20 million acres. *Solenopsis invicta* became the South's dominant ant and continued to spread throughout the rest of the century. Complaints about that insect's troubling habits multiplied. What had happened?

Several causes might account for this relatively slow spread and then sudden explosion. First, populations grow exponentially: one colony births two, each of those two more, and so on. In this way a small population can soon become huge. Under this scenario the ant population might have exploded after a lag time in which the insect was not much noticed. Second, in coming to North America, the red imported fire ant left behind predators and parasites that had hitherto controlled its population in South America.[70] This liberation may have accentuated *Solenopsis invicta*'s exponential growth. Third, what appeared to be a sudden explosion in the ant's population may have resulted, in part, from increased diligence in looking for it. During the initial period of its spread, when the insect caused fewer problems, the ant was easily ignored; only when its population grew enough for the insect to be considered a pest did entomologists systematically survey for the ant, which would have lead to finding more insects.

All three of these factors probably contributed to the ant's irruption, but there is also a fourth reason why the insect spread slowly and then irrupted: the changing ecology and economy of the South. When the black ant arrived in North America, Mobile was a city amid wilderness, not unlike Aveyros. Thick forests covered 80 percent of the land within a one hundred mile radius of Mobile.[71] An active port and the dynamism of urban life ensured that the city had plenty of open and disturbed habitats for the ants to occupy, but spreading beyond Mobile was difficult. Only a few roads and a languid river connected the city with the rest of the South. And beyond the forest wall stretched a

68. W. S. Creighton to M. S. Blum, 22 April 1968, Creighton papers.

69. H. G. Adkins, "The Imported Fire Ant in the Southern United States," *Annals of the Association of American Geographers* 60 (1970): 583.

70. S. D. Porter, D. F. Williams, R. S. Patterson, and H. G. Fowler, "Intercontinental Differences in the Abundance of Solenopsis Fire Ants (Hymenoptera: Formicidae): Escape from Natural Enemies?" *Environmental Entomology* 26 (1997): 373–84.

71. E. L. Ullman, "Mobile: Industrial Seaport and Trade Center" (Ph.D. diss., University of Chicago, 1942), 46.

landscape resistant to the fire ant's spread. Fragmented farms and what the USDA called "wastelands" (uncultivated farm land) covered the southern coastal plain, remnants of the Civil War. In the late-nineteenth-century South, poor whites and blacks had begun working small parcels of antebellum plantations, carving the huge, centralized tracts of land into a patchwork of small farms. These fragmented plantations anchored a new ecological regime: lush mats of vegetation—broom sedge, honeysuckle vines, cedar saplings, sassafras, sumac, and persimmon bushes—weaved amid split-rail fences and covered the wastelands. Quail, fox, raccoons, and rabbits bred and foraged in the brush. The brutal economic system that maintained the area's ecology stayed in place for three-quarters of a century, from the 1860s into the 1940s.[72] It seems reasonable to assume that the red imported fire ant did not spread very quickly during the first decades of its residence in North America because the southern environment proved relatively uninviting: the open and disturbed areas in which it survived best were present, but widely separated by areas of mature, relatively stable climax communities.

Beginning in the third decade of the twentieth century, though, growing in the 1930s and accelerating through the 1940s and 1950s, Mobile and the South underwent a social, demographic, and ecological revolution that helped *Solenopsis invicta* spread across the region and establish itself as the dominant ant species. The southern historian C. Vann Woodward called this upheaval the "bulldozer revolution."[73] While its roots run back to the 1920s, the revolution's watershed year was 1933, when Franklin Delano Roosevelt signed into law the Agricultural Adjustment Act (AAA). The AAA was intended to save southern farmers from the Great Depression by paying landlords to take their fields out of production, resulting in a decrease in supply that in turn would raise the price of crops and thus the revenue generated by farmers. Initially, southern landowners were reluctant to take the payouts, considering the program an assault on individual liberties.

72. This reconstruction is based on G. C. Fite, *Cotton Fields No More: Southern Agriculture, 1865–1980* (Lexington: University Press of Kentucky, 1984); P. Daniel, *Breaking the Land: The Transformation of Cotton, Tobacco, and Rice Cultures since 1880* (Urbana: University of Illinois Press, 1985); J. T. Kirby, *Rural Worlds Lost: The American South, 1920–1960* (Baton Rouge: Louisiana State University Press, 1987); B. J. Schulman, *From Cotton Belt to Sunbelt: Federal Policy, Economic Development, and the Transformation of the South, 1938–1980* (New York: Oxford University Press, 1991); A. E. Cowdrey, *This Land, This South: An Environmental History* (Lexington: University Press of Kentucky, 1996).

73. C. V. Woodward, *The Burden of Southern History* (Baton Rouge: Louisiana State University Press, 1993), 6.

Soon, though, they saw the advantage of accepting federal largesse and a flood of money from Washington, D.C. deluged the region. Landowners took the money, retired land, fired tenants, and, following the example of middle-class farmers who had experimented with Yankee methods in the 1920s, invested their capital in new technologies: tractors, harvesters, combines, and other mechanized agricultural tools commonly used in the Midwest and West but previously rare in the South.[74] Farm size grew as landowners knit previously fragmented farms into highly capitalized agricultural factories. Inefficient wastelands fell to the plow and split-rail fences gave way to straight ones, with no room for brush. Power and privilege in the South remained relatively concentrated in the same class before and after the bulldozer revolution, but standards of living increased dramatically as the South's economy became more productive and the sharecropping system that created and perpetuated poverty collapsed.[75]

With the onset of World War II, demand for Southern agricultural products skyrocketed and farm production grew accordingly. Tractors, harvesters, combines, and other such tools allowed the landowners to continue to increase the size of their farms. Between 1920 and 1969 the average size of an Alabama farm jumped from 76.4 acres to 188.3 acres. (Wastelands declined from 1.7 million acres in 1925 to 429,000 acres by 1964.) Similar changes remade rural areas throughout the South: the average size of a farm in Georgia, Louisiana, and Mississippi more than tripled between 1920 and 1969, to a point at which it was over two hundred acres.[76] New roads erased the distance between these farms and the city, allowing commerce to travel farther faster. "Old and New Alabama are merging," reported the WPA guide to Alabama, published in 1941. "Improved roads that reach even into hitherto inaccessible mountain regions are helping to free people from isolation."[77] Between 1925 and 1945, the number of Alabama farms within two-tenths of a mile of an all-weather road almost quintupled, from 27,000 to 137,000. Roads in Georgia grew at an even more astonishing rate: the same

74. On the mechanization of American farming, see D. Fitzgerald, *Every Farm a Factory: The Industrial Ideal in American Agriculture* (New Haven: Yale University Press, 2003).

75. For this history, see J. T. Kirby, *Rural Worlds Lost: The American South, 1920–1960* (Baton Rouge: Louisiana State University Press, 1987); T. Saloutos, *The American Farmer and the New Deal* (Ames: Iowa State University Press, 1982).

76. U.S. Bureau of the Census, *Census of Agriculture: 1959* (Washington, D.C.: Government Printing Office, 1961), 28:2, 32:2, 33:2, 35:2; U.S. Bureau of the Census, *Census of Agriculture: 1964* (Washington, D.C.: Government Printing Office, 1967), 28:7, 32:7, 33:7, 35:7; U.S. Bureau of the Census, *Census of Agriculture: 1969* (Washington, D.C.: Government Printing Office, 1971), 28:3, 32:3. 33:2, 35:2.

77. *Alabama: A Guide to the Deep South* (New York: Smith, 1941), 8.

statistic jumped from 9,000 in 1925 to 154,000 in 1945.[78] "Under clouds of dust the bulldozers are at work," wrote the author John Dos Passos. "Towed behind yellow tractors roadscrapers are leveling off pastureland. With a roar of motors mechanical shovels are chewing down red hills that a short time ago were woodlots of longleaf pines. . . . In two weeks a back country settlement with its shacks and barns and horse troughs and fences, all the frail machinery of production built up over the years by the plans and hopes and failures of generations of country people, will have vanished utterly."[79]

New residents inhabited this new landscape, and old denizens suffered. Soybeans replaced cotton as the South's leading agricultural product and the number of livestock rocketed as huge pastures were fenced off.[80] "This was the empire of the cotton baron," noted Dos Passos. "Now it's all going into stock farming and dairying."[81] The number of cattle in Alabama tripled between 1944 and 1967.[82] Meanwhile, the fate of other animals declined under the new regime: quail could not survive on farms and pastures shorn of brush, and their numbers plummeted.[83] "A bygone regime would look indulgently upon a sprawling rail fence with its innumerable elbows and pockets of wasteland," lamented hunter and English professor Havilah Babcock. "But modern agriculture, with its emphasis on maximum land utilization, must have fences that are straight, efficient, and characterless. The old rail contrivance, they said, was unscientific. Unscientific it was, and sometimes inefficient and wasteful. But it was the precise type of agriculture most congenial to the incidental production of game. Let's not forget that!"[84] In 1954, the Alabama Department of Conservation

---

78. U.S. Bureau of the Census, *Census of Agriculture: 1945* (Washington, D.C.: Government Printing Office, 1946), 17:4, 21:4.

79. J. Dos Passos, *State of the Nation* (Boston: Houghton Mifflin, 1944), 67. See also Twelve Southerners, *I'll Take my Stand: The South and the Agrarian Tradition* (New York: Harper and Brothers, 1930).

80. On soybeans, see H. D. Fornari, "The Big Change: Cotton to Soybeans," *Agricultural History* 53 (1979): 245–53.

81. J. Dos Passos, *State of the Nation* (Boston: Houghton Mifflin, 1944), 73.

82. Alabama Cattlemen's Association flier, Alabama Cattlemen's Association 1968 file, box 15, William Nichols papers, RG 194 (84-1), Auburn University Archives. On the southern cattle industry generally, see J. T. Kirby, *Rural Worlds Lost: The American South, 1920–1960* (Baton Rouge: Louisiana State University Press, 1987), 75, 340.

83. The complaints were ubiquitous, but for one statement see H. L. Stoddard, *The Cooperative Quail Study Association: May 11, 1931–May 1, 1943*, Tall Timbers Research Station, Miscellaneous Publications, no. 1 (Tallahassee: Tall Timbers Research Station, 1961), 248.

84. H. Babcock, *My Health Is Better in November* (Columbia: University of South Carolina Press, 1947), 77–78. See also J. H. Berryman, "Our Growing Need: A Place to Produce and Harvest Wildlife," *Journal of Wildlife Management* 21 (1957): 319–23.

admitted, "There may be small increases in quail from time to time, but the general trend is down."[85]

As the countryside changed, so did some forms of rural life. With diminishing space for wildlife, for example, quail hunting almost disappeared. Between 1962 and 2003, the number of quail hunters in Georgia dropped from 135,000 to 16,000.[86] A once-important ritual became a rarity. During the middle part of the twentieth century, the bird was the most popular game animal across much of the region and a source of regional identity for many.[87] Bobwhite quail, said Babcock, keeper of the quail's place in southern mythology, were "unreconstructed rebels," links to a time before the bulldozer revolution and before the Civil War, when southern culture was ascendant.[88] Quail hunters, said another commentator, represented "Truth with dirt on its face, Beauty with a briar scratch on its finger, Wisdom with Nature as its God, and the Hope of the future with goodwill towards man."[89] Hunting the birds reaffirmed a masculine code of honor: Babcock explained that when a hunter startles a covey of quail lurking in thick weeds, the birds burst from the cover with a sound reminiscent of "twenty pieces of dynamite" exploding.[90] The sound, said Babcock, "unmans me. I am for a moment disarmed, all my resolutions gone a-glimmering."[91] The heart raced, the stomach dropped. Nerves failed to fire. In an instant, without time to think, the hunter had to reconstitute his manhood, remember his arms, and regain his composure. He had to aim quickly, but shoot slowly. Any deviation from this pattern broke the gentleman's code: to fail to shoot was to remain unmanned; but to shoot uncontrollably was to remain in the grip of adrenaline. "I have to summon such Christian forbearance as I have," Babcock said,

85. Alabama Department of Conservation, *Report for the Fiscal Year October 1, 1953–September 30, 1954,* 100. See also "What Has Happened to Alabama's Quail?" Ford file, SG 9977, ALDoC papers.

86. R. Pavey, "Keeping the Quail Program Helps Preserve Proud Bird," *Augusta Chronicle,* 9 February 2003, C13. See also "Hunter Numbers on a Slide; Small Game Losing Its Appeal in Mississippi," *Commercial Appeal,* 6 October 1996, 13D.

87. F. S. Arant, *The Status of Game Birds and Mammals in Alabama* (Wetumpka, Ala.: Wetumpka Printing, 1939), 13.

88. H. Babcock, *I Don't Want to Shoot an Elephant* (New York: Holt, Rinehart and Winston, 1958), 1.

89. C. Dickey, "What Is a Quail Hunter?" *Florida Wildlife* 12 (October 1958): 5. For a fuller discussion, see S. A. Marks, *Southern Hunting in Black and White: Nature, History, and Ritual in a Carolina Community* (Princeton: Princeton University Press, 1991), 172–99, and the writings of Havilah Babcock.

90. H. Babcock, *My Health Is Better in November* (Columbia: University of South Carolina Press, 1947), 139.

91. Ibid., 225.

"to keep from overshooting them year after year."[92] As the abundance of quail decreased, some hunters struggled to hold onto the traditions that Babcock idealized, but others sought new rituals to give their lives meaning.[93]

Cities, as well, roiled from the combined effects of the New Deal and World War II. Roosevelt and the bloc of southern legislators preferentially sent defense funding to the South (and West), creating new industries that attracted dispossessed workers from rural areas.[94] Mobile's prewar population increased by almost 60 percent. Other cities underwent even more phenomenal growth: the population of Pascagoula, Mississippi, for example, swelled from four to thirty thousand during the war and the population of Panama City, Florida grew from twenty to sixty thousand.[95] In these cities, historian Pete Daniel wrote, "[w]orkers quickly filled available housing and then turned to house trailers, tents, hotels, conversions of all manner of structures into apartments, and the 'hot bed' solution of people working and using the bed on shifts. City services collapsed under the strain—reservoirs dried up, sanitation services and trash collection failed, public transportation ground under the weight of numbers, and schools ran in double shifts. . . . Places simply grew from town to city, with little attention to urban planning and zoning."[96] Adapting to this new demographic reality, Mobile annexed land at a prodigious rate (fig. 3). Parks, playgrounds, and schools sprang up to accommodate the Baby Boom generation.[97]

As the forest wall surrounding Mobile was cleared for the city's expansion, across the southern coastal plain the areas preferred by the imported fire ants proliferated: military bases, suburbs, cities, roadways, and cattle pastures, open and disturbed. In the 1930s, when the insects just started to stretch beyond the boundaries of Mobile, they mostly inhabited these kinds of areas: a fill strip along Highway 90 between

92. Ibid., 11.

93. For an example of the continued cultural power of hunting in the South, see T. Wolfe, *A Man in Full* (New York: Bantam Books, 1999), 9. The search for new rituals is discussed in S. Marks, *Southern Hunting in Black and White: Nature, History, and Ritual in a Carolina Community* (Princeton: Princeton University Press, 1991), 270–73.

94. D. W. Gantham, *The South in Modern America: A Region at Odds* (New York: Harper Collins Publishers, 1994), 173.

95. The population numbers are from J. T. Kirby, *Rural Worlds Lost: The American South, 1920–1960* (Baton Rouge: Louisiana State University Press, 1987), 304.

96. P. Daniel, *Standing at the Crossroads: Southern Life in the Twentieth Century* (New York: Hill and Wang, 1986), 136–37.

97. H. H. Jackson III, "Mobile since 1945," in *Mobile: The New History of Alabama's First City*, ed. M. V. R. Thomason (Tuscaloosa: University of Alabama Press, 2001), 279.

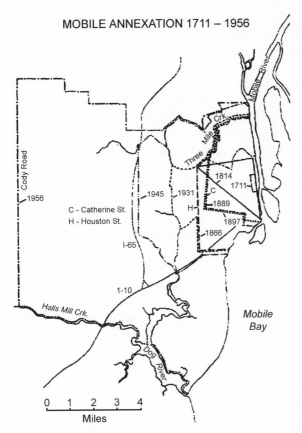

MOBILE ANNEXATION 1711 – 1956

Figure 3. This map traces the growth of Mobile, Alabama to 1956. The city expanded greatly in the years after 1931, felling the woods that surrounded it and linking with the rest of the South. The imported fire ant spread across the South in nursery shipments carried along the region's highway system. (From H. H. Jackson III, "Mobile since 1945," in *Mobile: The New History of Alabama's First City*, ed. M. V. R. Thomason [Tuscaloosa: University of Alabama Press, 2001], 282. Used by permission of the University of Alabama Press.)

Orange Grove, Mississippi and the Alabama state line, for instance, or the fields of Meridian, Mississippi, near a rail line that ran from Mobile.[98] And in the 1940s, entomologists continued to find the ants almost exclusively in these areas, one survey finding around seventy mounds per acre on grassy road strips and open fields, but only a few in dense woods or areas with thick undergrowth.[99] It only makes sense

---

98. M. R. Smith, "Report on Imported Fire Ant Investigations," 26 July 1949, Blum papers.

99. E. O. Wilson and J. H. Eads, "Special Report to the Alabama Department of Conservation," 16 July 1949, Blum papers. See also E. O. Wilson and J. H. Eads, "A Preliminary

that as the space devoted to these areas expanded, so did the ants' range—thus, a period of rather slow spread was followed by a large population explosion. The bulldozer revolution allowed *Solenopsis invicta* to irrupt.

The history of the Argentine ant's spread supports this conclusion. Its population irrupted in the 1930s, about three decades after the insect reached North America (fig. 4).[100] The fire ant irrupted only a decade or so after its introduction. Exponential growth, freedom from predators, and increasingly diligent surveys all probably played a part in this increase, but if these were the only factors, why did it take so long for the Argentine ant to irrupt? The reason is that exponential growth, liberation from predators, and better eyesight are not the only reasons for the spreads. Both species profited from the bulldozer revolution, too. Increased commerce helped them to spread, as southerners built the environments that the insects needed: houses filled with sweet foods for the Argentine ant and disturbed areas for the red imported fire ant. The Argentine ant population exploded a little earlier than the imported fire ant's, but that is to be expected since the Argentine ant lived in cities, and transportation routes connecting urban areas matured faster than those connecting the countryside, where the imported fire ant spread.

Yet despite the roadways, the open fields, and disturbed areas, the fire ant irruption could not have succeeded so well if not for one other innovation accompanying the bulldozer revolution: the growth of the nursery industry. Nurseries are urban agriculture, intensive sites of plant cultivation that grow along with cities. High demand for plants—to decorate suburban homes, to fill rights of way, to adorn the outside of skyscrapers— translates into high prices and nurseries survive, even thrive, in metropolitan areas where land prices drive out other kinds of agriculture.[101] For these reasons, nurseries had long persisted in Minnesota and New York and Illinois,

Report on the Fire Ant Survey," fire ant file, SG 9977, ALDoC papers; J. H. Stiles, R. H. Jones, and P. M. Dixon, "The Road Warriors: Spatial Patterns Associated with Corridor Use by the Red Imported Fire Ant," *Bulletin of the Ecological Society of America* 78 (supplement, 1997): 317; J. H. Stiles and R. H. Jones, "Distribution of the Red Imported Fire Ant, *Solenopsis invicta,* in Road and Powerline Habitats," *Landscape Ecology* 13 (1998): 335–46.

100. A. V. Suarez, D. A. Holway, and T. J. Case, "Patterns of Spread in Biological Invasions Dominated by Long-Distance Jump Dispersal: Insights from Argentine Ants," *Proceedings of the National Academy of Sciences* 98 (2001): 1097.

101. H. W. Lawrence, "The Geography of the U.S. Nursery Industry: Locational Change and Regional Specialization in the Production of Woody Ornamental Plants" (Ph.D. diss., University of Oregon, 1985), 68–77.

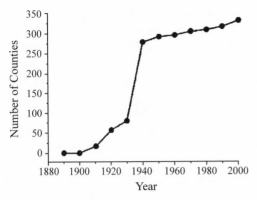

Figure 4. This chart shows the spread of the Argentine ant through U.S. counties. After its introduction, the insect spread only slowly, suddenly exploding in the 1930s, possibly because of the bulldozer revolution. The slower spread since the middle of the twentieth century may be due to the red imported fire ant, which also irrupted about this time and may have kept the Argentine ant in check.(From A. V. Suarez, D. A. Holway, and T. J. Case, "Patterns of Spread in Biological Invasions Dominated by Long-Distance Jump Dispersal: Insights from Argentine Ants," *Proceedings of the National Academy of Sciences* 98 [2001]: 1098. Copyright 2001 National Academy of Sciences, U.S.A.)

despite inhospitable climates. Railroads, too, had shifted power away from the South and toward the North and Midwest by charging discriminatory rates, favoring the North's big businesses over the poor, small nurseries south of the Mason-Dixon line.[102] With the coming of roads and trucks and cities in the South, however, the North's advantages dissipated and climate became a more decisive factor in the location of nurseries. Southern (and western) states, with mild climates and long growing seasons, seized the nursery industry. Shipping rhododendrons, azaleas, camellias, and chinaberry bushes, Mobile became the fifth largest nursery center in the country in 1950.[103]

The red imported fire ant spread across the South in burlap sacks and plastic containers filled with nursery stock, nestled amid the soil and plants' roots.[104] Under its own power, the insect spread only

102. G. Douglas, R. H. Jones, and P. Henegar, eds., *The History of the Southern Nurserymen's Association, Inc., 1899–1974* (Nashville: Southern Nurserymen's Association, 1974), 7.

103. H. W. Lawrence, "The Geography of the U.S. Nursery Industry: Locational Change and Regional Specialization in the Production of Woody Ornamental Plants" (Ph.D. diss., University of Oregon, 1985), 230–32.

104. On the ant's carriage in nursery stock, see M. R. Smith, "Report on Imported Fire Ant Investigations," 26 July 1949, Blum papers; W. G. Bruce, J. M. Coarsey, G. H. Culpepper, and C. C. Skipper, "Imported Fire Ant Survey," fire ant file, SG 9991, ALDoC papers.

about five miles per year.[105] But hidden in shipments of camellias and azaleas, for example, the red ant could travel hundreds of miles and establish beachheads in the open and disturbed habitats that it preferred: lawns, rights of way, and other nurseries. Over time, the ant backfilled the areas that it had skipped, consolidating its range. Comparison of a 1949 map (fig. 2) with a 1958 map (fig. 5) tells the tale. In 1949, a few ants, probably a mixture of black and red, had colonized northeast Mississippi and central Alabama, but many areas between their beachheads and Mobile were free of the insects. Eight years later, *Solenopsis invicta* had colonized those areas, too, and other red ants had leap-frogged to distant nurseries, continuing the process.

The Imported Fire Ants in the South

The tractors that graded land for pastures, plantations, military bases, housing developments, and shopping malls were not unlike the Río Paraguay and its tributaries, creating disturbance, making room for weedy, opportunistic species that could breed quickly, scavenge widely, and compete fiercely. Adapted to such a lifestyle, *Solenopsis invicta* followed the bulldozer across the South. Not only had it survived, it thrived, displacing other ants. The red species became the region's dominant ant, outpacing its black congener and even holding its own against the Argentine ant. A product of both the ant's biology and human actions, *Solenopsis invicta*'s irruption, in turn, affected both the southern environment and the people who lived there.

By the late 1940s and into the 1950s entomologists increasingly realized that the red imported fire ant was replacing the black. E. O. Wilson, a University of Alabama biology student who had become interested in ants as a high school student, and who later would become a Harvard professor and, after Wheeler, America's second great ant biologist, first noted the change in the fire ant population. He did not recognize the red and black forms as separate species—only different color variants of the same species—but saw that the red phase "replaced the dark phase; it is prevalent almost to the exclusion of the dark phase in the area of greatest spread, to [Alabama's] northeast; it is the form found by far in the greatest variety of situations."[106] No one knows exactly why the red ant won out over the black. Perhaps because

105. The five-mile-per-year average is reported in E. O. Wilson and W. L. Brown, "Recent Changes in the Introduced Population of the Fire Ant *Solenopsis saevissima* (Fr. Smith)," *Evolution* 12 (1958): 211. See also M. S. Blum to W. S. Creighton, 26 January 1969, Creighton papers.

106. For his earliest views, see the early fire ant notes, Wilson papers. The notes date back to 1948. The quote is from E. O. Wilson to W. S. Creighton, 28 February 1949, Creighton papers.

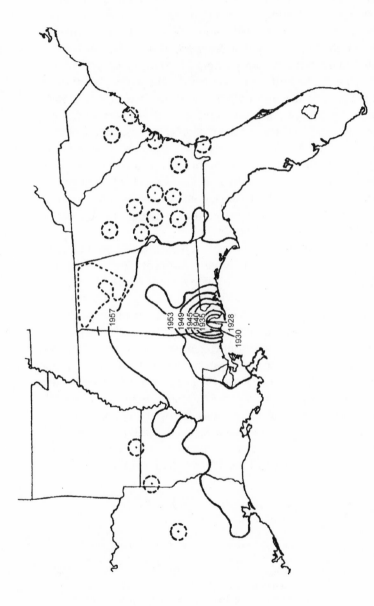

Figure 5. This map details the spread of the imported fire ant through 1957, although parts of it are somewhat misleading. First, the ranges from 1928 to 1949 are hypothesized. No one monitored the spread of the insects at the time. Second, the map combines the movement of *Solenopsis richteri* and *Solenopsis invicta*. The important point to note, though, is the large expansion between 1953 and 1957. By the early 1950s, the USDA and others monitored the insects fairly closely, so these estimates are reasonably good. The changing southern landscape, spread of roads, and growth of the nursery industry may have fueled this explosion.(From E. O. Wilson and W. L. Brown, "Recent Changes in the Population of the Fire Ant *Solenopsis saevissima* [Fr. Smith]," *Evolution* 12 [1958]: 212. Used by permission of *Evolution*.)

it is a better competitor. It also seems reasonable to assume that the red ant may have had an advantage coming to North America late. Reaching American shores as the bulldozer revolution began, it could rapidly and easily expand, colonizing new areas before the black ant had a chance, and then overwhelming areas that *Solenopsis richteri* had once held. Whatever the cause of the red ant's triumph over the black, Wilson was later able to quantify the shift with his Harvard colleague William Brown: 80 percent of the imported fire ants living along Beach Road in Gulf State Park, Alabama, for example, had been black in 1949; eight years later, less than 40 percent were black, the rest red.[107] Similar increases in the red fire ant's population—and declines in the black—occurred across most of the South. *Solenopsis richteri,* though, did not completely succumb. Although many entomologists were not aware of it, the species held on in northeast Mississippi and northwest Alabama, even as *Solenopsis invicta* spread across the rest of the South and replaced the black imported fire ant in southern Alabama and southern Mississippi.[108]

The red imported fire ant displaced other ants as well. *Solenopsis geminata,* most likely present in North America for centuries, gave way, as did *Solenopsis xyloni,* the southern fire ant, a native of North America.[109] Other native ants also lost ground to the immigrant, and even the Argentine ant seems to have suffered in some competitions with the red imported fire ant. Stagnation followed *Linepithema humile*'s population explosion in the 1930s, and some scientists suggest that *Solenopsis invicta* curbed the Argentine ant's spread: the red fire ant population had irrupted and the insect was no longer overwhelmed by the Argentine ant or easily driven to extinction, as perhaps had been the case in New Orleans.[110] "The fire ant has apparently reduced populations of the Argentine ant . . . to such a degree that this one-time major pest of the household is of little or no importance in Alabama," wrote the entomologist Frank Arant in the 1970s.[111] Two decades later, ento-

107. E. O. Wilson and W. L. Brown, "Recent Changes in the Introduced Population of the Fire Ant *Solenopsis saevissima* (Fr. Smith)," *Evolution* 12 (1958): 213.

108. S. W. Taber, *Fire Ants* (College Station: Texas A&M Press, 2000), 58–60. H. B. Green was aware of the black ant's survival. See H. B. Green to M. S. Blum, 29 April 1968, Creighton papers.

109. E. O. Wilson and W. L. Brown, "Recent Changes in the Introduced Population of the Fire Ant *Solenopsis saevissima* (Fr. Smith)," *Evolution* 12 (1958): 217–18. See also W. R. Tschinkel, "Distribution of the Fire Ants *Solenopsis invicta* and *S. geminata* (Hymenoptera: Formicidae) in Northern Florida in Relation to Habitat and Disturbance," *Annals of the Entomological Society of America* 81 (1988): 76–81.

110. A. V. Suarez, D. A. Holway, and T. J. Case, "Patterns of Spread in Biological Invasions Dominated by Long-Distance Jump Dispersal: Insights from Argentine Ants," *Proceedings of the National Academy of Sciences* 98 (2001): 1097.

111. F. S. Arant, "Notes on the Imported Fire Ant," n.d., Blum papers.

mologists estimated that *Solenopsis invicta* composed up to 97 percent of the ant fauna in the open and disturbed areas that it preferred.[112]

This shift in the composition of the ant fauna had repercussions for the ecology of the South that even now are only beginning to be explored. Ants rival worms in the amount of earth that they move while building mounds and digging tunnels.[113] They alter the chemistry of soils, concentrating some elements in their mounds, impoverishing nearby areas of others.[114] And their presence favors some species over others. Aphids, for example, do well around many ant species because the ants protect them in order to lap up a sugar secreted by the aphids. Most kinds of solitary insects, on the other hand, often do poorly in the presence of ants.[115] Ants, in some respects, are comparable to humans, both species that can dramatically change the environment to fit their needs with repercussions that are not always easily predictable.[116] The red imported fire ant, alien to North America, altered the ecology of its new home, as do most other imported species.[117] The intensity and importance of these changes would be a matter of contention during the fire ant wars, and scientists have not yet unraveled the cascade of consequences that followed the red imported fire ant's invasion. There is, however, little doubt that the ant affected crops, animal life, and humans.

The opportunistic red ant ate whatever was available—nursery stock, crops, or insects.[118] In 1949, Wilson and James Eads, another University of Alabama biology student, estimated that each year the imported fire ant caused $70,000 worth of damage to soybean crops in Baldwin County, Alabama, alone.[119] Farmers complained that the insect ate germinating seeds, potatoes, corn, and a host of other crops. The year of Wilson and Eads' survey, one Alabaman described her garden after the ant attacked it: "*Not one* carrot ever came up after *many* plantings, neither one stalk of bell pepper. I also planted a *high stack* of flower seeds,

---

112. S. D. Porter, H. G. Fowler, and W. P. Mackay, "Fire Ant Mound Densities in the United States and Brazil (Hymenoptera: Formicidae)," *Journal of Economic Entomology* 85 (1992): 1158.

113. E. O. Wilson, *The Insect Societies* (Cambridge: Harvard University Press, 1971), 1.

114. S. W. Taber, *Fire Ants* (College Station: Texas A&M Press, 2000), 20.

115. B. Hölldobler and E. O. Wilson, *Journey to the Ants: A Story of Scientific Exploration* (Cambridge: Harvard University Press, 1994), 8.

116. E. O. Wilson, "The Evolutionary Significance of the Social Insects," in *Insects, Science and Society,* ed. D. Pimentel (New York: Academic Press, 1975), 25–31; E. O. Wilson, *Success and Dominance in Ecosystems: The Case of the Social Insects* (Oldendorf: Ecology Institute, 1990).

117. On the dangers of alien species, see *Harmful Non-Indigenous Species in the United States* (Washington, D.C.: Office of Technology Assessment, 1993).

118. M. R. Smith, "Report on Imported Fire Ant Investigations," 26 July 1949, Blum papers.

119. E. O. Wilson and J. H. Eads, "A Preliminary Report on the Fire Ant Survey," fire ant file, SG 9977, ALDoC papers.

and planted them right—not too deep—and not one came up. Then I asked my husband to please plant another package of petunia seeds, they wouldn't come up for me, and he did it carefully . . . but the ants got *every one* of them also." Adding insult to injury, the ant stung her feet whenever she tried to work in her garden.[120] Tractors, too, had difficulties negotiating infested fields. The ant builds huge mounds that bake hard in the southern sun. Tractor blades could break or the spinning metal could toss a mass of angry insects onto the driver's back. In some areas, almost one hundred mounds peppered a single field, making the use of tractors impossible.[121]

The ant also attacked a wide range of animals, both wild and domestic. Its presence reduced the population of lizards, for example, and the insect ate quail just hatching from eggs as well as other birds.[122] A swarm could devour a half-pound rat in a couple of hours and fish nipping at alates that accidentally landed in water were sometimes stung to death.[123] The ant worried sheep and cows and reduced the population of deer.[124] "The ants crawl into the fawn's nose and its stomach," while the animal instinctively freezes in the presence of a predator, noted P. C. Hanes, a Texas animal rehabilitator, in 1988. "When we wash out their stomachs, we find hundreds of fire ants."[125] The invader also devoured boll weevils, sugar cane borers, and other insects, its opportunistic habits leading the red ant to eat both animals valued by humans and animals considered pests.[126] The importance of these attacks, though, was not

120. Mrs. W. D. Monk to B. E. Thomas, 20 January 1949, fire ant infestation file, SG 9998, ALDoC papers. Emphasis in original.

121. G. H. Culpepper, "Damage of Imported Fire Ants to Certain Agricultural Interests of Dallas County Alabama," fire ant file, SG 17020, ALDoC papers.

122. For an overview, see C. R. Allen, S. Demarais, and R. S. Lutz, "Red Imported Fire Ant Impact on Wildlife: An Overview," *Texas Journal of Science* 46 (1994): 51–59. For the relationship between the fire ant and lizards see R. H. Mount, S. E. Trauth, and W. H. Mason, "Predation by the Red Imported Fire Ant, *Solenopsis invicta* (Hymenoptera: Formicidae), on Eggs of the Lizard *Cnemidophorus sexlineatus* (Squamata: Teiidae)," *Journal of the Alabama Academy of Sciences* 52 (1981): 66–70; R. H. Mount, "The Red Imported Fire Ant, *Solenopsis invicta* (Hymenoptera: Formicidae), as a Possible Serious Predator of Some Native Southeastern Vertebrates: Direct Observations and Subjective Impressions." *Journal of the Alabama Academy of Sciences* 52 (1981): 71–78.

123. On the ant eating a rat, see M. J. Killion and S. B. Vinson, "Ants with Attitudes," *Wildlife Conservation* 98 (January–February 1995), 48. On fish deaths, see C. Contreras and A. Labay, "Rainbow Trout Kills Induced by Fire Ant Ingestion," *Texas Journal of Science* 51 (1999): 199–200.

124. On deer, see C. R. Allen, R. S. Lutz, and S. Demarais, "Ecological Effects of the Invasive Nonindigenous Ant, *Solenopsis invicta*, on Native Vertebrates: The Wheels on the Bus," *Transactions of the 63rd North American Wildlife and Natural Resources Conference*, 1998: 56–65.

125. E. Yoffe, "The Ants from Hell," *Texas Monthly* 16 (August 1988): 82.

126. T. E. Reagan, "Beneficial Aspects of the Imported Fire Ant: A Field Ecology Approach," in *Fire Ants and Leaf-Cutting Ants: Biology and Management*, ed. C. S. Lofgren and R. K. Vander Meer (Boulder: Westview, 1986), 58–72.

clear in the late 1940s and remains open to question today. In 1949, for example, Wilson told a group of hunters, "They of course destroy wildlife. How much they destroy wildlife I cannot say. It seems that wildlife is being decimated in the main areas of infestation and the ant is the main contributing factor toward their decimation."[127] He noted privately, though, that the claim that ants devoured vertebrates in any significant number "needs a *lot* of checking" and neither he nor USDA entomologists found evidence that the ant attacked wildlife during their surveys in the 1940s.[128] Some fifty years later, one wildlife biologist insisted that tales of the ant attacking quail were more of a red herring than a real problem, but others disagreed.[129]

There was no doubt, however, that the ant stung people, and stung painfully. In the 1940s, Mississippi potato harvesters refused to go into infested fields, and Wilson heard that other laborers did, too.[130] Whenever field hands neared the colonies, a swarm of ants attacked them; a single laborer could suffer scores of painful stings, and these could, in rare cases, be fatal. The 1984 death of former Alabama state senator Finis St. John III was attributed to fire ant stings that he received while driving his tractor, and in the 1990s *Solenopsis invicta* killed a number of nursing home residents.[131] An eighty-seven-year-old woman with Alzheimer's disease died a day after the insects stung her 1,625 times. As the ant invaded new areas, outdoor recreation became difficult: the ant bothered picnickers and scared students playing in schoolyards.[132] It stung President George Bush in 1992 when he was

127. Meeting of Advisory Board, 17 February 1949, SG 17017, ALDoC papers.

128. The quotation is from the early fire ant notes, Wilson papers. For Wilson and the USDA's failure to document attacks on wildlife, see E. O. Wilson and J. H. Eads, "Special Report to the Alabama Department of Conservation," 16 July 1949; M. R. Smith, "Report on Imported Fire Ant Investigations," 26 July 1949, both in Blum papers; see also W. G. Bruce, J. M. Coarsey Jr., M. R. Smith, and G. H. Culpepper, "Survey of the Imported Fire Ant, *Solenopsis saevissima* var. *richteri*, Special Report S-15, July 1949, fire ant file, SG 9998, ALDoC papers.

129. L. A. Brennan, "Fire Ants and Northern Bobwhites: A Real Problem or a Red Herring," *Wildlife Society Bulletin* 21 (1993): 351–54; C. R. Allen, R. D. Willey, P. E. Myers, P. M. Horton, and J. Buffa, "Impact of Red Imported Fire Ant Infestation on Northern Bobwhite Quail Abundance Trends in Southeastern United States," *Journal of Agricultural and Urban Entomology* 17 (2000): 43–51.

130. Meeting of Advisory Board, 17 February 1949, SG 17017, ALDoC papers; M. R. Smith, "Report on Imported Fire Ant Investigations," 26 July 1949, Blum papers.

131. On the senator, see B. Harrell, "Mounds of Trouble from the Fire Ant: When Southerners Talk about the Red Scourge, They Don't Mean Russia," *Atlanta Journal-Constitution*, 6 July 1986, M10. On the nursing home residents, see R. D. deShazo, D. F. Williams, and E. S. Moak, "Fire Ant Attacks on Residents in Health Care Facilities: A Report of Two Cases," *Annals of Internal Medicine* 131 (1999): 424–29.

132. For a review of problems faced by people enjoying the outdoors, see R. T. Ervin and W. T. Tennant Jr., "Red Imported Fire Ants' *(Solenopsis invicta)* Impact on Texas Outdoor Recreation," in *Applied Myrmecology: A World Perspective*, eds. R. K. Vander Meer, K. Jaffe, and A. Cedeno

bird hunting.[133] In the words of a Texas journalist, "Fire ants have changed Texans' relationship with nature. The earth beneath our feet has been transformed into teeming cities of venom-filled stingers."[134]

Conclusion

Pests are not born; they are made. "No species is inherently a pest," wrote the wildlife biologist and environmental philosopher Aldo Leopold, "and any species may become one."[135] Evolving on a South American flood plain, the red imported fire ant adapted to life in open and disturbed habitats. It thrived in fields exposed to the sun, using its powerful sting to keep competitors at bay; and the insect spread through disturbed areas, its catholic diet allowing *Solenopsis invicta* to colonize areas where food sources were not ensured. It traveled well and competed fiercely, but the ant was not innately a pest; it only became a pest in the right environment.

When the ant first journeyed to North America, it did not initially find such a hospitable environment. While parts of Mobile, Alabama did provide open and disturbed habitats of the sort that the ant favored, the rest of the South was not so easy to invade. Although *Solenopsis invicta* had left behind predators and parasites, it spread slowly. By the 1940s the red fire ant was replacing its black cousin, but neither had moved far from Mobile. *Solenopsis invicta* did not irrupt until the 1950s.

The timing of the red fire ant's population explosion illustrates the major thesis of this chapter—and this book: that natural history and human history intertwine. The imported fire ant originated long before the first human paced the earth, and for eons it lived far from the naked primate; but, over time, the ant's fate became bound to the history of some *Homo sapiens*. The insect stowed away aboard ships sailing from South America; it may have become established in Mobile, Alabama because of another insect that people accidentally introduced into New Orleans; and it spread along human-made avenues to the open and disturbed habitats that Americans created in abundance. Once the new transportation routes and new habitats were in place, the ant's biology took over: its weediness allowed *Solenopsis invicta* to spread exponentially. The ant entered both the southern ecosystem and southern culture.

(Boulder: Westview, 1990), 504–10. Picnickers' hazards are described in E. Yoffe, "The Ants from Hell," *Texas Monthly* 16 (August 1988): 82.

133. "Bush Still Nursing Wounds after Tangle with Fire Ants," *Dallas Morning News,* 4 January 1992, 15A.

134. E. Yoffe, "The Ants from Hell," *Texas Monthly* 16 (August 1988): 82.

135. A. Leopold, *The River of the Mother of God and Other Essays,* ed. S. L. Flader and J. B. Callicott (Madison: University of Wisconsin Press, 1991), 309.

Chapter Two: Grins
a Prohibitive Fracture,
1945–1957

> But between us and the Insects,
> namely nine-tenths of the living, there grins a prohibitive fracture
> empathy cannot transgress: (What Saint made a friend of a roach or
> preached to an ant-hill?) Unrosed by a shame, unendorsed by a sorrow,
> blank to a fear of failure, they daunt alike the believer's
> faith in a fatherly providence and the atheist's dogma of purely
> random events.
>
> W. H. Auden, "The Aliens"

On April 13, 1587, in the St. Julien's region of the country that would become France, the Catholic Church put on trial some weevils that had been attacking the grapes of local vintners.[1] (The trial's conclusion is lost to history, the relevant pages of the transcript eaten by insects.)[2] Three and a third centuries later, on December 11, 1919, the citizens of Enterprise, Alabama dedicated a statue to another weevil, the cotton boll weevil, "in profound appreciation [of what it] has done as a herald of prosperity."[3] These two events span not only time and distance, but also illustrate the continuum of possible responses to invading insects: they can be vilified or celebrated—or treated in a way that falls somewhere in between.

Thus, when the imported fire ant irrupted across the southern tier of the United States, Americans had a range of possible responses from which to choose. The insect did eat crops and quail, but these depredations were not an automatic justification for hating the bugs. Residents of Enterprise celebrated the boll weevil although it destroyed cotton crops because the insect also forced the city to diversify and strengthen its economy.[4] Proposing counterfactuals can be dangerous, but it is not inconceivable that Americans could have praised the imported fire ant for illustrating that, while the bulldozer revolution

1. E. P. Evans, *The Criminal Prosecution and Capital Punishment of Animals* (London: Faber and Faber, [1906] 1978), 41.

2. Ibid., 49.

3. Committee on the Future Role of Pesticides in U.S. Agriculture, *The Future Role of Pesticides in US Agriculture* (Washington, D.C., National Academy Press, 2000), 35.

4. On the history of the boll weevil, see J. D. Helms, "Just Lookin' for a Home: The Cotton Boll Weevil and the South" (Ph.D. diss., University of Florida, 1977).

improved the southern standard of living, the changes came with a cost: weedy, stinging, species replaced those such as quail, which were much preferred. Americans, though, at least initially, chose to treat the insect as invaders that needed to be fought and destroyed. Why? Why did the ant become a villain? Why did it become the subject of the largest, longest, most costly insect eradication program in American history?

Two answers have been offered in the past. One blames the insect: the U.S. Department of Agriculture (USDA) argued that the ant was such a pest that it made its own destruction necessary.[5] The second answer comes from, among others, historians Pete Daniel and Emily Shore: the USDA invented the menace, they say, for its own selfish reason; the ant itself was a mere nuisance.[6] Neither of these answers does justice to the complexity of the situation, though. The ant did cause problems, but it took several large steps to go from merely complaining about the insect to attempting to control it and, finally, to deciding to eradicate it from the country. The ant's troublesome habits cannot fully explain why those steps were taken. And yet the USDA cannot be given all the credit—or blame—for creating the ant's image. The department was not the first to investigate the insect. A decade before the eradication program began, a host of entomologists and a newspaper reporter observed the ant. This early work laid the groundwork for the USDA's policy, which found favor among some groups in the South, indicating at least some acceptance of the USDA's position among those who lived with the insect. The ant's identity did not result solely from its biology or from the USDA's description. Rather, its image emerged from an interaction between the insect and several groups of people—the USDA, the Alabama Department of Conservation, entomologists, hunters, and journalists. Only when all of these came together did the ant become the "blackest of villains" and a worthy target of an eradication program.[7]

5. See, for example, L. J. Padget, "Some Facts about the Imported Fire Ant Program and Reported Wildlife Losses," 20 June 1958, imported fire ant-wildlife losses file, box 198, Plant Pest Control Division papers, RG 463, National Archives and Records Administration, College Park, Md. (hereafter PPC papers).

6. P. Daniel, "A Rogue Bureaucracy: The USDA Fire Ant Campaign of the Late 1950s," *Agricultural History* 64 (1990): 99–114; E. F. Shores, "The Red Imported Fire Ant: Mythology and Public Policy, 1957–1992," *Arkansas Historical Quarterly* 53 (1994): 320–39; P. Daniel, *Lost Revolutions: The South in the 1950s* (Chapel Hill: University of North Carolina Press, 2000), 78–86.

7. Quote from R. H. Allen, "History of the Imported Fire Ant in the Southeast," *Proceedings of the Twelfth Annual Conference of the Southeastern Association of Game and Fish Managers,* October 1959, 228.

The Meaning of Ants

By the time that *Solenopsis invicta* reached North America, ants had been a favorite object of moral lessons for millennia, "statements as to [their] intelligence, diligence, avarice, foresight and policy . . . a part of the common patrimony that we acquire in prepatory school," the Nobel Prize–winning playwright Maurice Maeterlinck said.[8] "Go to the ant, you sluggard," advised the author of Proverbs (6:6), for instance. "Consider its way, and be wise! It has no commander, no overseer or ruler, yet it stores its provisions in summer and gathers its food at harvest." During the nineteenth century and into the twentieth, however, changing agricultural practices—especially in the U.S. Midwest—and a generation of self-consciously secular ant biologists combined to create a new image of the insects, one that did not replace the traditional notion, but grew along side and challenged it. Ants became less a paragon of virtue and more of a "monster," in the original sense of that word: an omen of bad things to come.

In the second half of the nineteenth century, agriculture in the American Midwest intensified. Railroads and canals connected markets, increasing demand for produce and the availability of capital. New agricultural machines became available and farm size grew. Crops were planted in dense, homogenous stands. (The South would follow this same trajectory in later years, although prodded by the federal government's investment, not market forces.)[9] These farms were bonanzas for insects, providing copious amounts of food in relatively restricted areas, and throughout the later years of the nineteenth century insect populations irrupted across the region: locusts, the chinch bug, and the Colorado potato beetle, for example, all devoured crops.[10] Prior generations—although sometimes vociferously battling insects— often had looked indulgently upon insects. Nettlesome though they may have been, bugs were also seen as products of God's imagination, material proof of His being. The rise and growth of large-scale com-

8. M. Maeterlinck, *The Life of the Ant,* trans. B. Miall (New York: John Day, 1930), 3.

9. J. T. Kirby, *Rural Worlds Lost: The American South 1920–1960* (Baton Rouge: Louisiana State University Press, 1987), 1–22.

10. For the connection between changing agricultural practices and insect irruptions, see C. Sorenson, "The Rise of Government Sponsored Applied Entomology, 1840–1870," *Agricultural History* 62 (1988): 104. For locusts, see J. T. Schlebecker, "Grasshoppers in American Agricultural History," *Agricultural History* 27 (1953): 85–93; for the chinch bug, see W. C. Sorenson, *Brethren of the Net: American Entomology, 1840–1880* (Tuscaloosa: University of Alabama Press, 1995), 116–20; for the potato beetle, see R. A. Casagrand, "The Colorado Potato Beetle: 125 Years of Mismanagement," *Bulletin of the Entomological Society of America* 33 (1987): 142–50.

mercial farming eroded some of this belief.[11] Many now saw insects as humanity's competitors; they occupied the far side of a prohibitive fracture: "Something in the insects," Maeterlinck wrote on another occasion, "seems to be alien to the habits, morals and psychology of our globe, as if it had come from some other planet, more monstrous, more energetic, more insensate, more atrocious, more infernal than our own."[12] The USDA's Bureau of Entomology, established in 1877 to combat a locust outbreak, led America's war against insects, its growth providing a rough index of the growing hostility toward bugs: between 1897 and 1912 it increased from 21 staffers to 339.[13] In so far as ants were insects, they, too, suffered this decline in reputation, and were increasingly seen as antagonistic to human civilization.

The reevaluation of ants also results from changed interests, methods, and goals on the part of those who studied the insects. Beginning in the nineteenth century, a generation of self-consciously secular ant biologists, or "myrmecologists," as they called themselves, elbowed aside religious interpreters of ant life (although pastors and clerics continued to study the insects for a time).[14] In America, William Morton Wheeler took up this line of work, becoming the world's leading myrmecologist from the late 1910s until his death in 1937, and training America's second generation of myrmecologists, including William Creighton. Widely read in sociology, philosophy, and the humanities, as well as biology—the philosopher Alfred North Whitehead reportedly quipped that the myrmecologist was the only man "who would have

11. W. C. Sorenson, *Brethren of the Net: American Entomology, 1840–1880* (Tuscaloosa: University of Alabama Press, 1995), 92–126.

12. For a discussion of the antagonism toward insects common at this time, see T. R. Dunlap, *DDT: Scientists, Citizens, and Public Policy* (Princeton: Princeton University Press, 1981), 17–38. Maeterlinck's quotation is from W. M. Wheeler, *The Social Insects: Their Origin and Evolution* (London: Keegan Paul, Trench, Trubner, 1928), 321.

13. M. W. Rossiter, "The Organization of the Agricultural Sciences," in *The Organization of Knowledge in Modern America, 1860–1920*, eds. A. Oleson and J. Voss (Baltimore: Johns Hopkins University, 1979), 218. The USDA had an entomologist on staff since the 1850s, but the growth of the Bureau of Entomology really began with the creation of the U.S. Entomological Commission in 1877. See R. W. Dexter, "The Organization and Work of the U.S. Entomological Commission, 1877–1882," *Melsheimer Entomological Series* 26 (1979): 28–32.

14. For religious interpretations of ant life, see H. C. McCook, *The Honey Ants of the Garden of the Gods, and the Occident Ants of the American Plains* (Philadelphia: Lippincott, 1887); E. Wasmann, *Instinct und Intelligenz im Thierreich: Ein Kritischer Beitrag zur Modernen Thierpsychologie* (Freiburg im Breisgau: Herder, 1899). For the secularization of myrmecology, see A. Forel, *Out of My Life and Work*, trans. B. Miall (London: George Allen and Unwin, 1937), 25 and passim; M. A. Evans and H. E. Evans, *William Morton Wheeler, Biologist* (Cambridge: Harvard University Press, 1970), 168; J. F. M. Clark, " 'The Ants Were Duly Visited': Making Sense of John Lubbock, Scientific Naturalism, and the Senses of Social Insects," *British Journal for the History of Science* 30 (1997): 151–76.

been both worthy and able to sustain a conversation with Aristotle"—
Wheeler was concerned with the social disruption wrought by the birth
of the industrial order in America and thought that a study of ants,
apart from its intrinsic interest, might shed light on his country's transition from Republic to polyglot, capitalistic, modern society.[15] He concentrated his study on ant social life, and drew parallels between the
insects' life history and human civilization.

The key to ant sociology, Wheeler concluded, was tropholaxis, the
sharing of food. In a colony, only a few ants gather food, which they
then pass on to the rest of the nest. Whereas the Jesuit scholar Erich
Wasmann interpreted this sharing in moral terms, Wheeler shrugged
off Wasmann's conclusions as Catholic casuistry, arguing that moral
imperatives did not drive this process and insisting on a material, scientific explanation for the behavior.[16] Ants, he said, were driven to
share by instincts that had evolved over time. They were therefore "the
most thoroughgoing of communists, to whom individual possession as
such has no meaning beyond its benefit to the community as a
whole."[17] The communism of ants had its positive aspects: the ants'
tendency to share made them among the most successful creatures on
earth. But the corollaries of this form of social organization were troubling. In the fetid darkness of many ant nests, mites scurried and the
maggoty young of beetles squirmed. Following their communist instinct to share, ants fed these interlopers, directing vital energy away
from their larvae and to the parasites. Heavily parasitized colonies died.
Wheeler had studied ants from a materialistic perspective and found
not moral exemplars, but creatures on the verge of decadence.

And humans flirted with a similar fate, he warned. Both humans
and ants were social creatures that had developed through similar evolutionary phases and seemed to follow the same trajectory, but ants
had developed social life earlier and so had reached senescence first.
Humans, however, were not far behind. Modern civilization hinted at
the decadence of human societies. Wheeler wrote, "We not only tolerate
but even foster in our midst whole parasitic trades, institutions, castes
and nations, hordes of bureaucrats, grafting politicians, middlemen,

15. On Wheeler, see M. A. Evans and H. E. Evans, *William Morton Wheeler, Biologist* (Cambridge: Harvard University Press, 1970). The quotation is from G. H. Parker, "William Morton
Wheeler, 1865–1937," *Biographical Memoirs of the Members of the National Academy of Sciences*
19 (1938): 220. For fears of modernity, see T. J. J. Lears, *No Place of Grace: Antimodernism and the
Transformation of American Culture, 1880–1920* (New York: Pantheon Books, 1981).

16. M. A. Evans and H. E. Evans, *William Morton Wheeler, Biologist* (Cambridge: Harvard University Press, 1970), 169.

17. W. M. Wheeler, "Notes about Ants and Their Resemblance to Man," *National Geographic*
23 (1912): 747.

profiteers and usurers, a vast and varied assortment of criminals, hoboes, defectives, prostitutes, white-slavers and other purveyors of anti-social proclivities, in a word so many non-productive, food-consuming and space-occupying parasites that their support absorbs nearly all the energy of the independent members of society."[18] Cooperation, Wheeler acknowledged, is an awesome force for good—cooperation promised to counter the horrors of nationalism so graphically illustrated during World War I—but cooperation could be perverted, too, and lead to social disintegration. Ants were the most successful insect group on earth, a status that Wheeler attributed to their highly developed social organization; but they were also monsters, warnings of what could happen if humans subsumed too much of their individuality into society. As the poet W. H. Auden wrote decades later, what insects "mean to themselves or to God is a meaningless question: they to us are quite simply what we must never become."[19]

Broadly understood as a rebuke to the religious understandings of ants, Wheeler's ideas leaped the fence surrounding the Ivory Tower, entered the popular culture, and merged with the increasing distaste of insects generally. His view found a place in Arpaud Ferenczy's 1924 novel *The Ants of Timothy Thümmel*, for example—in which a tribe of African ants suffered when politicians, soldiers, and clerics decided that they need not work, but should instead be served by others—and the popular accounts of Maeterlinck and the British scientist Julian Huxley. In these works, fear of communism's consequences mixed with admiration for the ant's cooperative lifestyle.[20] By the mid to late 1930s, though, and continuing into the 1940s, writings about ants became increasingly pessimistic, dovetailing with critiques of Soviet communism and Nazi totalitarianism. "Is there anything subliminally foreboding about this fascination with the anthill?" the editors of *Christian Century* asked about the popularity of ant farms in postwar America. "Is it a dreadful future we see through this translucent plastic—regimented automatons, driven, dutiful in their prescribed pointless doing?"[21] Sparked by the work of the psychologist T. C. Schnierla, the American press developed a taste for stories about army ants—South American insects that moved as a phalanx across the jungle, leaving a path of

18. W. M. Wheeler, *Social Life among the Insects* (New York: Harcourt, Brace, 1923), 198.

19. W. H. Auden, "The Aliens," in *Epistle to a Godson and Other Poems* (New York: Random House, 1972), 33.

20. A. Ferenczy, *The Ants of Timothy Thümmel* (New York: Harcourt, Brace, 1924); M. Maeterlinck, *The Life of the Ant*, trans. B. Miall (New York: John Day, 1930); J. Huxley, *Ants* (New York: Jonathan Cape and Harrison Smith, 1930).

21. "Of Ants and Men," *Christian Century* 74 (13 November 1957): 1339.

devastation.[22] Swarming killers of the jungle, the *Saturday Evening Post* called them.[23] "Their conduct seems often a grotesque parody of man at his worst," *Newsweek* reported. "The ants bivouac regularly, like Caesar's legions; as camp followers they have plump parasitic beetles; they can even claim a sort of airforce of flies and scavenger birds."[24] The myrmecologist Caryl Haskins caught the mood in 1939 with *Of Ants and Men* (a selection of the month by the Scientific Book Club) and later in 1951 with *Of Societies and Men,* arguing that ants lived in total-itarian states, the insect equivalents of Nazis and communists.[25]

This view of ants found its most refined expression in Carl Stephenson's 1938 story "Leiningen versus the Ants," and the story's popularity over the next two decades proves the degree to which ants were seen as dangerous: William Conrad voiced Leiningen in a 1948 radio play and Charlton Heston brought the character to the silver screen six years later.[26] Originally published in *Esquire* magazine, then selling about half a million copies each month, the tale chronicled the battle between Leiningen, a white plantation owner in South America, and "twenty square miles of life destroying ants."[27] An examplar of civ-ilization, Leiningen had struggled to carve his plantation out of the oppressive Amazonian jungle and he stood his ground when army ants marched toward his property, telling a Brazilian official, "When I began this model farm and plantation three years ago, I took into account all that could conceivably happen to it. And now I'm ready for anything and everything—including your ants."[28] The fight was epic: he filled a moat with water, but the ants crossed on rafts of leaves; he shoveled dirt on the advancing hordes, but they continued unimpeded; he built

22. "Go to the Army Ant," *Newsweek* 31 (12 April 1948): 54; A. Devoe, "The World of Ants," *American Mercury* 69 (August 1949): 225–29; T. C. Schnierla, "Army Ants," *Scientific American* 178 (1948): 16–23; "All about Army Ants," *Newsweek* 40 (28 July 1952): 53; B. Eddy, "Go to the Ants, and Be Warned," *New York Times Magazine,* 12 December 1948, 22, discusses parasol ants, a dif-ferent group, but the language was similar to that used in discussions of army ants.

23. J. O'Reilly, "Swarming Killers of the Jungle: Army Ants," *Saturday Evening Post* 225 (16 May 1953): 36, 177–82.

24. "March of the Ants," *Newsweek* 49 (18 March 1957): 84.

25. C. P. Haskins, *Of Ants and Men* (New York: Prentice-Hall, 1939); C. P. Haskins, *Of Societies and Men* (New York: Viking, 1951). J. S. W., "Of Ants and Men," *Journal of the New York Ento-mological Society* 48 (1940): 102, notes that Haskin's earlier book was featured by the book club.

26. C. Stephenson, "Leiningen versus the Ants," *Esquire* 10 (December 1938): 98–99, 235–41. For a reprint, see "Leiningen versus the Ants," *Senior Scholastic* 52 (5 April 1948): 25–26. I found a copy of the radio play in the Multnomah County Library, where it was part of the collection "Escape," vol. 2; *The Naked Jungle,* Paramount Pictures, 1954.

27. C. Stephenson, "Leiningen versus the Ants," *Esquire* 10 (December 1938): 99. The data on the magazine's distribution is from *The Sixth New Year* (Chicago: Esquire, 1938), 13.

28. C. Stephenson, "Leiningen versus the Ants," *Esquire* 10 (December 1938): 98.

a wall of fire around his land, but it ran out of fuel. The ants devoured his crops and forced Leiningen and his servants to take refuge on a small disc of raised land. Finally, Leiningen wrapped himself in kerosene-soaked rags and dashed two miles across a sea of ants that could strip a buffalo to bones "before you can spit three times" to open a dam and drown the insects.[29] The ants ate through the rags and gobbled his flesh; he staggered and fell unconscious, but not before he flooded his plantation, destroying everything that he had built—and killing the ants. The story illustrated the mettle, ingenuity, and determination of humans in the face of adversity; it also emphasized the threat posed by ants. They did not deserve praise, but opprobrium—not emulation, but hatred. The proper response was not to observe and celebrate them, but to kill them. The ants were a people, as Proverbs 30:25 said, but a dangerous people, epitomizing the mindless savagery that Americans saw in Nazism, communism, and other forms of totalitarianism.[30]

Fire Ants in the United States

Ant biology and the changing conditions of American society combined to tilt popular perception of the insects away from the idea that ants were paragons of virtue and toward the view that they were despicable. Thus, when the imported fire ants, first *Solenopsis richteri* and then *Solenopsis invicta,* reached North America and spread across the South, the imported insects confronted a culture pre-adapted to dislike them. A broad distaste for ants does not fully explain the response to the red fire ants, however, especially since that distaste was not monolithic: some still found ants admirable, worthy of study, and worthy of contemplation. How, then, did the red imported fire ant become a villain? Who studied it? How did they approach it? And what did they see?

As it spread across the countryside, the red imported fire ant brought a great deal of attention to itself. It ruined crops and its mounds appeared seemingly out of nowhere, rendering the use of tractors impossible. It crawled across schoolyards, across military bases, and across parks, stinging people and scaring farm hands. It worried livestock. Letters of complaint reached officials in Alabama and Mississippi. "The Fire Ants are taking the whole neighborhood here," complained one man in Fairhope,

29. Ibid.

30. By the time of the red imported fire ant's irruption, the idea of totalitarianism encompassed both Nazism and communism, however ideologically different those movements were. See L. K. Adler and T. G. Paterson, "Red Fascism: The Merger of Nazi Germany and Soviet Russia in the American Image of Totalitarianism, 1930's to 1950's," *American Historical Review* 75 (1970): 1046–64.

Alabama. "We have tried so many remedies with no results."[31] State governments took notice. In 1948, the state gave Mississippi State University entomologist Clay Lyle and his graduate student Irma Fortune $15,000 to study the ant and recommend controls.[32] School children surveyed the state to determine the extent of the invasion.[33] A year later, the Alabama Department of Conservation paid for a study of the insect by E. O. Wilson and James H. Eads and purchased some insecticides to control it; shortly thereafter, the Alabama Polytechnic Institute developed an interest in the ant.[34] The states pressed the USDA for help, but the department dragged its heels.[35] Eventually, though, the USDA also started some investigations. These studies were one of three strands that, braided together, came to form the image of the imported fire ant.

These early studies were small, underfunded, and staffed by a high proportion of young scientists—Wilson and Eads were undergraduates (Wilson was only twenty), Fortune a graduate student, and the USDA's Leyburn Lewis only a year out of graduate school—probably because they worked cheap and study of the ant was not considered very prestigious.[36] The scientists had few resources. To make do with what they had, the entomologists spent most of their time gathering and corroborating reports from farmers, since the farmers could direct the entomologists to the ant—which was easier than scouring every acre in the South.[37] Talking with farmers and studying their fields reinforced

31. J. P. Hermecz to the Alabama Department of Conservation, 16 May 1949, fire ant infestation file, SG 9998, Alabama Department of Conservation files, director's administrative files, Alabama Department of Archives and History, Montgomery, Ala. (hereafter ALDoC papers).

32. C. Lyle and I. Fortune, "Notes on an Imported Fire Ant," *Journal of Economic Entomology* 41 (1948): 834.

33. Early fire ant notes, in E. O. Wilson papers, Library of Congress, Washington, D.C. (hereafter Wilson papers).

34. On Alabama Polytechnic Institute's developing interest, see the handwritten note on the back of E. O. Wilson and J. H. Eads, "A Preliminary Report on the Fire Ant Survey," SG 9977, ALDoC papers.

35. For the USDA's dilatory response, see A. S. Hoyt to B. E. Thomas, 8 March 1949; B. E. Thomas to A. S. Hoyt, 14 March 1949, both in fire ant infestation file, SG 9998, ALDoC papers.

36. For Lewis's biography, see *American Men of Science: A Biographical Dictionary*, 11th ed. (New York: Bowker, 1966).

37. For examples, see I. Fortune, "The Biology and Control of *Solenopsis saevissima* variety *richteri* Forel," (Master's thesis, Mississippi State College, 1948); E. O. Wilson and J. H. Eads field notes, SG 9977, ALDoC papers; State survey, n.d., Ford file, SG 9977, ALDoC papers; "State Survey of Fire Ants under Way," *Alabama Conservation* 20 (April–May 1949): 4; M. R. Smith, "Report on Imported Fire Ant Investigations," 26 July 1949, Murray S. Blum papers, in author's possession (hereafter Blum papers); W. G. Bruce, J. M. Coarsey Jr., M. R. Smith, and G. H. Culpepper, "Survey of the Imported Fire Ant, *Solenopsis saevissima* var. *richteri*," Special Report S-15, July 1949, fire ant file, SG 9998, ALDoC papers; E. O. Wilson to P. Charam, 14 July 1964, fire ant file, box 247, PPC papers.

the image that the ant was a pest. Fortune concluded that the mounds accounted for most of the damage done by the ant, a point made by others.[38] She found that a hayfield could become unmowable six months after invasion by the ant. The entomologists also discovered that the insect ate crops, including cabbage, okra, eggplant, and potato. Wilson and Eads estimated that the ant caused "a total crop damage in Mobile County of $177,610; and in Baldwin County of $357,612."[39] Even when farmers told Fortune that the ant sometimes ate other pest insects and therefore might be beneficial, she dismissed the claim since she had already seen so many problems caused by the ant that easily outweighed its positive aspects.[40]

Irma Fortune and USDA entomologists also sought ways to control the ant. The department opened a laboratory in Spring Hill, Alabama, where Leyburn Lewis studied the insect. He raised the ant in the laboratory and found a fungus that grew on its eggs—a biological control, he thought. But it proved useless.[41] Lewis also experimented with cultural control, that is, the adoption of farming techniques that minimize the damage done by insects. When he heard that the ant seemed to avoid a certain kind of corn, Lewis planted it in the laboratory garden, only to observe that over time the corn proved not to be resistant to the insect.[42] The entomologists, however, spent most of their time studying a third method of control, insecticides, maybe because biological and cultural control did not work, but also because chemicals were revolutionizing entomology. Allied forces had used DDT during World War II to protect soldiers from typhus and malaria.[43] In the years after the war, many entomologists thought that the chemical and others like it could be used to protect American agriculture and public health.[44] These insecticides had a low acute toxicity—especially when

38. I. Fortune, "The Biology and Control of *Solenopsis saevissima* variety *richteri* Forel," (Master's thesis, Mississippi State College, 1948), i, 6–7.

39. E. O. Wilson and J. H. Eads, "A Report on the Imported Fire Ant *Solenopsis saevissima* var. *richteri* Forel in Alabama," 16 July 1949, 1, Blum papers.

40. I. Fortune, "The Biology and Control of *Solenopsis saevissima* variety *richteri* Forel," (Master's thesis, Mississippi State College, 1948), 9.

41. On Lewis's work with the fungus, see L. F. Lewis, "The Imported Fire Ant—Investigations in the Biology of the Imported Fire Ant, *Solenopsis saevissima* var. *richteri*," SG 17020, ALDoC papers.

42. L. F. Lewis, "The Imported Fire Ant—Investigations in the Biology of the Imported Fire Ant, *Solenopsis saevissima* var. *richteri*," SG 17020, ALDoC papers.

43. On entomologists in the war, see E. C. Cushing, *History of Entomology in World War II* (Washington, D.C: Smithsonian Institution, 1957).

44. On entomologists' hopes for the postwar years, see E. N. Cory, "Entomology and the War," *Journal of Economic Entomology* 36 (1943): 355–58; P. N. Annand, "The War and the Future of Entomology," *Journal of Economic Entomology* 37 (1944): 1–9; R. E. Campbell, "What Do I Get

compared to the arsenicals and cyanide salts used in years past—were cheap, and killed efficiently. Many entomologists came to see their science as a form of chemistry.[45] They turned away from studies of natural history, bionomics, and taxonomy: cultural control and biological control seemed passé, as did the biological studies needed to make them work. One entomologist said, "With the advent of DDT and the many new insecticides . . . that followed in quick succession, laboratory and field testing of new pesticides increased to a point where it dominated all other activity."[46] As the power of the new insecticides became increasingly clear, some entomologists imagined a new agenda for their science. They could eradicate insects. "Is not this an auspicious time for entomologists to launch determined campaigns for the complete extermination of some of the pests which have plagued man through the ages?" Clay Lyle asked in a celebrated 1946 speech to the American Association of Economic Entomologists.[47]

Lyle's student, Irma Fortune, studied the effectiveness of a number of different chemical insecticides, concluding that DDT and its relative, chlordane, were "markedly effective," as were several other chemicals.[48] USDA entomologists also had good luck with chlordane. The eradication ideal had not yet permeated the department, but the scientists were so confident that the insecticide could control the ant that they voluntarily closed the fire ant laboratory in 1953, recommending that infested areas be treated with two pounds of chlordane per acre of cultivated land, or four pounds per acre on uncultivated land.[49] The

---

for My Three Dollar Membership?" *Journal of Economic Entomology* 45 (1952): 1–4. On DDT's passage from military technology to domestic use, see E. P. Russell III, "The Strange Career of DDT: Experts, Federal Capacity, and Environmentalism in World War II," *Technology and Culture* 40 (1999): 770–96.

45. For the developing connection between entomology and chemistry, see P. Palladino, *Entomology, Ecology and Agriculture: The Making of Careers in North America, 1885–1985* (The Netherlands: Harwood Academic Publishers, 1996), 38–41. See also R. Van den Bosch, *The Pesticide Conspiracy* (New York: Doubleday, 1978), 21; M. Kogan, "Integrated Pest Management: Historical Perspectives and Contemporary Developments," *Annual Review of Entomology* 43 (1998): 244.

46. G. C. Decker, "Don't Let the Insects Rule," *Journal of Agricultural and Food Chemistry* 6 (1958): 101.

47. C. Lyle, "Achievements and Possibilities in Pest Eradication," *Journal of Economic Entomology* 40 (1947): 1.

48. I. Fortune, "The Biology and Control of *Solenopsis saevissima* variety *richteri* Forel," (Master's thesis, Mississippi State College, 1948), 77.

49. Work Project I-h-8: "Develop Methods of Control of the Imported Fire Ant," September 1953, Entomology Research Division: Records of Insects Affecting Man and Animals Branch, Miscellaneous, 1935–1964, Agricultural Research Service papers, RG 310, National Archives and Records Administration, College Park, Md. (hereafter ARS papers).

poison could be applied to the ant nests—most effectively if the mounds were disturbed first—or worked into the soil with fertilizer. Demonstrations were given throughout the region, showing the proper method to apply the chemical. The insecticide gave one year's worth of control and cost less than one cent per mound.[50]

But, despite the evidence that the ant was destructive and the insecticides potent, not everyone agreed on the severity of the problems caused by the insect nor that it would be wise to eradicate it. Fortune thought that the ant was a major pest that needed to be destroyed. Wilson concurred.[51] Everyone knew that the native fire ants, *Solenopsis geminata* and *Solenopsis xyloni,* damaged crops and sometimes ate wildlife—the Mississippi entomologist Marion Smith called *Solenopsis xyloni* "one of the worst ant pests."[52] It seemed likely that the imported fire ant was at least as damaging as these insects.[53] Scientists at the Alabama Polytechnic Institute, however, could only bring themselves to deem the import a "minor pest" that had "become a more important economic pest" as it spread.[54] Meanwhile, Department of Agriculture entomologists admitted, "Efforts to run down reported serious damage to crops usually resulted in finding a gross exaggeration of damage"; and other claims of the ant's depredations were unusual rather than instructive. Fortune, for example, heard that the ant had devoured twenty acres of corn in George County, but noted, "This is the only occurrence of their attacking a corn crop. They evidently resort to this when other food is scarce, since they are normally meat or grease-eating ants."[55] In July 1949, after Fortune, Wilson, and Eads

50. Work Project I-h-8: "Develop Methods of Control of Fire Ants," fire ant file, SG 17020, ALDoC papers.

51. I. Fortune, "The Biology and Control of *Solenopsis saevissima* variety *richteri* Forel," (Master's thesis, Mississippi State College, 1948), 9; Meeting of the advisory board, 17 February 1949, SG 17017, ALDoC papers.

52. S. W. Clark, "The Fire Ant," *Texas Citriculture* 1929: 26; E. O. Essig, "The Fire Ant," *Texas Citriculture* 1930: 15; S. W. Clark, "The Control of Fire Ants in the Lower Rio Grande Valley," *Bulletin of the Texas Agricultural Experiment Station* 435 (1931): 1–12; M. R. Smith, "Consideration of the Fire Ant *Solenopsis xyloni* as an Important Southern Pest," *Journal of Economic Entomology* 29 (1936): 120–22; A. Mallis, "The California Fire Ant and Its Control," *Pan-Pacific Entomologist* 14 (1938): 87–91; M. R. Smith, *House-Infesting Ants of the Eastern United States: Their Recognition, Biology, and Economic Importance* (Washington, D.C.: USDA, 1965), 39 (quotation).

53. On the connections drawn between the natives and the imported fire ant, see M. R. Smith, "Report on Imported Fire Ant Investigations," 26 July 1949, Blum papers; W. G. Bruce, J. M. Coarsey Jr., M. R. Smith, and G. H. Culpepper, "Survey of the Imported Fire Ant, *Solenopsis saevissima* var. *richteri*," Special Report S-15, July 1949, fire ant file, SG 9998, ALDoC papers.

54. W. G. Eden and F. S. Arant, "Control of the Imported Fire Ant in Alabama," *Journal of Economic Entomology* 42 (1949): 976.

55. I. Fortune, "The Biology and Control of *Solenopsis saevissima* variety *richteri* Forel," (Master's thesis, Mississippi State College, 1948), 7.

had concluded their work, entomologists with the USDA concluded, "The imported fire ant is considered to be more of a nuisance than a destroyer of plant and animal life."[56] The ant needed to be controlled, but the department's myrmecologist, Marion Smith, felt that southerners should learn to live with the insect. "It is my candid opinion that although the imported fire ant may be eradicated in small, easily treated areas, it will never be eradicated from the large general area it now occupies," he wrote. A report co-authored by Smith and three other entomologists reinforced this opinion. "It is very doubtful if the imported fire ant will ever be eradicated from the United States," they said.[57]

E. O. Wilson and the Imported Fire Ant

Aside from his investigation of the fire ant study for the state of Alabama, Wilson contributed a second strand to what would become the braided image of the fire ant. As he began his career, Wilson straddled the divide between chemically and biologically oriented entomology. Early on, he dreamed of the day that he would "ride around in one of those green pick up trucks used by the U.S. Department of Agriculture's extension service to visit rural areas" and help farmers with their bug problems.[58] He got a chance to see what that kind of work would be like when he studied the imported fire ant; he also learned the value of insecticides. In 1948 Wilson told a group of hunters, "I think it would be futile to try to find a natural parasite. I think to attain total eradication of a large area or reduce the number of ants to a negligible number, you must develop a new technique," and recommended that they hire a chemist.[59] At the same time, Wilson, under the tutelage of his biology professor Ralph Chermock, learned that entomology could also be a vehicle for answering biological questions.[60] He developed an enthusiasm for evolutionary theory, and by the time he graduated Wilson no longer wanted to work for the

56. W. G. Bruce, J. M. Coarsey Jr., M. R. Smith, and G. H. Culpepper, "Survey of the Imported Fire Ant, *Solenopsis saevissima* var. *richteri*," Special Report S-15, July 1949, fire ant file, SG 9998, ALDoC papers.

57. M. R. Smith, "Report on Imported Fire Ant Investigations," 26 July 1949, Blum papers; W. G. Bruce, J. M. Coarsey Jr., M. R. Smith, and G. H. Culpepper, "Survey of the Imported Fire Ant, *Solenopsis saevissima* var. *richteri*," Special Report S-15, July 1949, fire ant file, SG 9998, ALDoC papers.

58. E. O. Wilson, "In the Queendom of Ants: A Brief Autobiography," in *Leaders in the Study of Animal Behavior: Autobiographical Perspectives*, ed. D. A. Dewsbury (Lewisburg, Pa.: Bucknell University Press, 1985), 465.

59. Meeting of the advisory board, 17 February 1949, SG 17017, ALDoC papers.

60. E. O. Wilson, *Naturalist* (Washington, D.C.: Island Press, 1994), 107–15.

USDA. Instead he went to the University of Tennessee for graduate study, then, at the prodding of William Creighton and William Brown, who was already at Harvard, to the school where Wheeler had worked and Creighton had studied.[61] In 1963, Wilson repudiated his early interest, calling the entomological practices that had developed around insecticides "a closed world of second-rate science."[62] But, even as he made this transition, Wilson confirmed his suspicion that the red imported fire ant was "one of the worst ant pests in the Western Hemisphere."[63] The red imported fire ant, he concluded, was a mutated version of the ant that Henry Walter Bates had seen attacking the village of Aveyros in the Amazon basin—the blackest of villains, indeed.

While exploring evolutionary theory, Wilson became, as he later remembered, "enchanted by the idea of dominant animals and the succession of dynasties," his interest probably sparked by reading the paleontologist William Diller Matthew's *Climate and Evolution*.[64] Matthew argued that northern latitudes were the center of evolution. Made strong by selection in a harsh environment, creatures from the north pushed south, displacing primitive and less successful species. Over the course of history, several waves of dominant northern groups had spread across the globe, forcing weaker animals to take refuge in the tropics, where, Matthew said, the mild climate was more forgiving and allowed poor competitors to survive. Studying the imported fire ant, Wilson thought that he saw this pattern inverted: a fire ant from the South was displacing its northern relatives; especially impressive, the red ant was displacing not only fire ants native to North America, but the black form of the imported fire ant, as well.[65] Intrigued by what he saw, Wilson opened a new line of study to investigate the reason for

61. J. B. Buhs, "Building on Bedrock: William Steel Creighton and the Reformation of Ant Systematics," *Journal of the History of Biology* 33 (2000): 48.

62. "Statement Read before the U.S. Senate Subcommittee on Reorganization and International Organizations," 8 October 1963, William Steel Creighton papers, an unprocessed box of material incorporated in the uncatalogued E. O. Wilson papers, Library of Congress, Washington, D.C. (hereafter Creighton papers).

63. Meeting of the advisory board, 17 February 1949, SG 17017, ALDoC papers.

64. E. O. Wilson, *Naturalist* (Washington, D.C.: Island Press, 1994), 212; E. O. Wilson, "Variation and Adaptation in the Imported Fire Ant," *Evolution* 5 (1951): 74. On Matthew see W. D. Matthew, *Climate and Evolution* (New York: New York Academy of Sciences, [1915] 1939); E. H. Colbert, *William Diller Matthew, Paleontologist: The Splendid Drama Observed* (New York: Columbia University Press, 1992). Context for Matthew's work can be found in P. J. Bowler, *Life's Splendid Drama: Evolutionary Biology and the Reconstruction of Life's Ancestry 1860–1940* (Chicago: University of Chicago, 1996).

65. E. O. Wilson, "Variation and Adaptation in the Imported Fire Ant," *Evolution* 5 (1951): 74.

the red ant's dominance.[66] Why, he asked, was the red ant spreading so rapidly?

Wilson conducted a series of experiments to try to decipher the reason for the red ant's success. He raised colonies of both the red and black ants, varying the temperature and the amount of food available to see if environmental factors caused the differences between the two color forms, since ants were known to develop differently in different environments.[67] But, regardless of the conditions, red colonies remained red and black colonies remained black. Wilson also transplanted red ant larvae into nests of the black fire ant and *Solenopsis geminata* to see if the something in the nest might determine the ant's development.[68] The larvae, though, developed the color of their birth mother, not their adopted colony. These results led Wilson to conclude that the difference between the two ants was genetic. But the difference was not stark: the ants were not separate species; rather, the red ant was "a favorable mutation introduced into the population."[69] About ten years earlier, the geneticist Theodosius Dobzhansky had defined evolution as a change in gene frequency—an epochal transformation in the understanding of evolution, under which the kind of large-scale processes that Matthew described were understood to result from the slow accumulation of mutant genes, each with small effects, that overtime changed the genetic composition of a population.[70] Studying the imported fire ant, then, Wilson was on the vanguard of biology, observing evolution in action, as ants with one set of genes replaced another, a miniature version of the pattern that Matthew had described.[71]

The find was exciting and helped Wilson to make a name for himself in the small world of myrmecology. He turned his studies into a master's thesis at the University of Alabama, published a paper in the scientific journal *Evolution,* and extended the idea in a series of investigations of other ants and other organisms.[72] Wilson's discovery also

66. On Wilson's developing interest, see Meeting of the advisory board, 17 February 1949, SG 17017, ALDoC papers; E. O. Wilson to W. S. Creighton, 28 February 1949, Creighton papers.

67. E. O. Wilson, "Variation and Adaptation in the Imported Fire Ant," *Evolution* 5 (1951): 74.

68. Ibid.

69. Ibid., 79.

70. T. Dobzhansky, *Genetics and the Origin of Species* (New York: Columbia University Press, 1937), 11.

71. E. O. Wilson, "The Fire Ant," *Scientific American* 198 (1958): 39.

72. Wilson discusses his graduate work in E. O. Wilson to W. S. Creighton, 20 February 1951, Creighton papers. It was published as E. O. Wilson, "Variation and Adaptation in the Imported Fire Ant," *Evolution* 5 (1951): 68–79. He extended his ideas in E. O. Wilson, "Adaptive Shift and Dispersal in a Tropical Ant," *Evolution* 13 (1959): 122–44; E. O. Wilson, "Some Ecological Characteristics of Ants in New Guinea Rain Forests," *Ecology* 40 (1959): 437–47; E. O. Wilson,

added a new concern about the spread of the imported fire ant. While at the University of Tennessee, he studied a collection of South American fire ants for his (second) master's thesis, discovering that the red ant lived there also, but did not seem as destructive in South America as in the American South.[73] Wilson speculated that the red ant reached North America sometime after the black ant, and something about its mutated genome made it a better competitor than its black relative. Since the red ant was better adapted to life in North America than its congener, and even better than native ants, Wilson expected it to spread widely and cause problems. Summarizing his ideas in a 1958 article, he wrote, "Energized by the highly adaptive genes of the [red] form, its rapidly growing populations may within the next 10 or 20 years come to cover all of the southeastern states. As a warm temperate-zone species that nests exclusively out-of-doors, it may never succeed in pushing north much beyond its present limits in Alabama and North Carolina, but within this range it will undoubtedly continue to wax as one of the most noxious of all insect pests."[74]

Making the ant into a fierce mutant, Wilson added to the insect's notoriety; he also did so by tightening the connection between the insect invading North America and the one that Bates had seen in the Amazonian jungle. When Wilson began his intensive investigation of the insect, it was classified as *Solenopsis saevissima richteri* (Richter's savage fire ant)—a subspecies of the ant that Bates had seen, differentiated from the insect that attacked Aveyros by the stripe across its gaster.[75] Wilson initially accepted this classification, but as he became more familiar with ant taxonomy, he questioned why the insect was considered a separate subspecies.[76]

"The Nature of the Taxon Cycle in the Melanesian Ant Fauna," *American Naturalist* 85 (1961): 169–93; E. O. Wilson, "The Challenge from Related Species," in *The Genetics of Colonizing Species,* ed. H. G. Baker and G. L. Stebbins (New York: Academic Press, 1965), 7–27; R. H. MacArthur and E. O. Wilson, *The Theory of Island Biogeography* (Princeton: Princeton University Press, 1967).

73. The results of Wilson's studies are presented in E. O. Wilson, "O Complexo *Solenopsis saevissima* na America do Sul (Hymenoptera: Formicidae)," *Memorio do Instituto Oswald Cruz* 50 (1952): 49—68; E. O. Wilson, "Origin of the Variation in the Imported Fire Ant," *Evolution* 7 (1953): 68–79; E. O. Wilson to W. S. Creighton, 28 February 1949; W. S. Creighton to M. R. Smith, 23 October 1957, both in Creighton papers.

74. E. O. Wilson, "The Fire Ant," *Scientific American* 198 (1958): 41.

75. Actually, at the very beginning of Wilson's studies, the ant was considered a variety of Bates's ant, called *Solenopsis saevissima saevissima* var. *richteri.* But in 1950, shortly after Wilson began his studies, Creighton changed it to *Solenopsis saevissima richteri,* making the ant a subspecies, part of a general reworking of ant taxonomy that eliminated the varietal category altogether. See J. B. Buhs, "Building on Bedrock: William Steel Creighton and the Reformation of Ant Systematics," *Journal of the History of Biology* 33 (2000): 43–53.

76. E. O. Wilson to W. S. Creighton, 20 February 1951, Creighton papers.

Wilson's skepticism about the taxonomy of the imported fire ant was part of a broader disenchantment with the concept of the subspecies.[77] At Harvard, he and William Brown studied how taxonomists divided groups into subspecies, and concluded the process was arbitrary. "[T]here ain't any such thing" as a subspecies, Brown said. They made the point in two scientific papers that claimed subspecies were figments of the classifying imagination and criticized other taxonomists for being unscientific.[78] Their critique was part of a more general pattern of professional behavior by Wilson and Brown: they were young, ambitious, and looking for a way to wedge themselves into the scientific community. As Wheeler had criticized his intellectual forerunners (and as Creighton had revolted against some of Wheeler's ideas), Wilson and Brown attacked their elders, distinguishing their work from that of others. Wilson, for example, said that Creighton's taxonomic work was built on "a shifting and treacherous foundation."[79] And Brown compared Creighton to a gatekeeper blocking his entry into the myrmecological fraternity: "Your letters *in toto* put me in mind of a brush I had with a customs officer at Naples (lovely fellow)," Brown wrote. "He accused me of all sorts of evil intentions, made me disrobe in part, stole my cigarettes under some obscure bureaucratic license, cursed at me . . . and peppered me with a stiff and steady monsoon of garlic. I nevertheless forbore . . . because I very much wanted to get ashore in that town. I can also put up with almost anything from W. S. Creighton."[80] Doing away with the subspecies category provided a firm foothold for Wilson and Brown in myrmecological circles: it proved the weakness of previous taxonomy and established them as leaders in the field.[81]

In some ways, the decision to reject the subspecies category seems a tempest in a teapot. But, as the biologist Raymond Pearl said, taxonomy is "the bricks with which the whole structure of biological knowledge

77. The beginning of Wilson's disenchantment with the idea of the subspecies is documented in E. O. Wilson to A. C. Cole, 29 March 1952, Arthur Charles Cole papers, an unprocessed box of material incorporated in the uncatalogued E. O. Wilson papers, Library of Congress, Washington, D.C.

78. The quotation is from W. L. Brown to W. S. Creighton, 22 May 1953, Creighton papers. The two scientific papers are E. O. Wilson and W. L. Brown, "The Subspecies Concept and Its Taxonomic Application," *Systematic Zoology* 2 (1953): 97–111; W. L. Brown and E. O. Wilson, "The Case against the Trinomen," *Systematic Zoology* 3 (1954): 174–76.

79. E. O. Wilson to R. E. Gregg, 27 November 1954, Creighton file, Frank Morton Carpenter papers, HUG (FP) 4264.10, Harvard University Archives (hereafter Carpenter papers).

80. W. L. Brown to W. S. Creighton, 29 October 1954, Creighton file, Carpenter papers.

81. J. B. Buhs, "Building on Bedrock: William Steel Creighton and the Reformation of Ant Systematics," *Journal of the History of Biology* 33 (2000): 55; E. O. Wilson, *Naturalist* (Washington, D.C.: Island University Press, 1994), 204–8.

is reared."[82] By dividing and naming groups of plants and animals, tax-onomists organize nature, and in so doing provide a framework for others to think about it. The imported fire ant was no longer a relative of the ant that Bates had seen a century earlier in the Amazon basin, but the same species. Wilson told the USDA that Bates's observation proved that the ant could be a pest, and a noxious one.[83] The imported fire ant, though, was not just the same ant that Bates had seen attacking Aveyros. It was a mutated form of that ant, adapted to spread across North America as no other fire ant could. Studying the ant from the perspective of evolutionary biology and taxonomy, while ambitiously seeking to make his mark in the field, Wilson added to the widespread belief that the ant was a destructive and terrible pest.

Operation Ant

The third strand that would be braided into the image of the imported fire ant came not from entomologists, either chemically or biologically inclined, but from a hunter and newspaper columnist, Bill Ziebach. Ziebach had been reporting on the dwindling quail population in the wake of the bulldozer revolution when Alabama farmer Silas Evans told the journalist that he had found the real cause of the quail's decline: ants. Evans invited Ziebach to his home in Malcolm, where the two men walked the property, Evans pointing out one anthill after another. The insect, Evans told Ziebach, bothered his family and his animals. A neighbor reported that he that he could not "pick peas because they are covered with ants" and that he had to "keep the children in the house or be with them when they play in the yard. Every time an ant stings, it gives them a fever." Most troubling to Ziebach was Evans's statement that the "moment a quail egg 'pips' the ants enter and completely consume the chick."[84]

Others had accused fire ants of killing quail in the past.[85] In the 1930s, the wildlife biologist Herbert Stoddard witnessed a species of fire ant killing about 15 percent of the bobwhite that he was trying to raise on hunting plantations in Georgia and Florida.[86] "This is one of the most pitiful forms of destruction encountered in the study," Stoddard wrote in *The Bobwhite Quail*, a popular manual for cultivating

82. R. Pearl, "Trends of Modern Biology," *Science* 56 (1922): 582.

83. E. O. Wilson to P. Charam, 14 July 1964, fire ant file, box 247, PPC papers.

84. B. Ziebach, "Deadly 'Fire Ants' Infest Wide Area near Mobile," *Mobile Press-Register*, 2 May 1948, 13B.

85. J. T. Emlen Jr., "Fire Ants Attacking California Quail Chicks," *Condor* 40 (1938): 85–86.

86. H. L. Stoddard, *The Bobwhite Quail: Its Habits, Preservation and Increase* (New York: Scribner, 1931), 193.

the bird, "for the helpless baby bobwhites are literally eaten alive and squirming, with most of the flesh eaten off on one side of the face and vertebræ."[87] Stoddard, though, eventually concluded that however disgusting the ant's attack seemed, the insect was not a major threat to quail. Ziebach, on the contrary, thought that the ant was responsible for the bird's decline—an attack not only on southern nature, but on southern culture. As Ziebach surveyed the farm, he saw nests rising above the cropped grass of cow pastures and dotting the fields. They lined highway rights of way. Particularly hard hit areas suffered under the weight of as many as one hundred mounds per acre; other areas, however, had no nests. On average, Ziebach estimated, fifty mounds covered each acre. Multiplying by the forty square miles that he had examined with Evans, Ziebach calculated that there were over 6 million ant nests near Malcolm, Alabama alone. Although he had no idea how many ants lived in a mound, certainly there had to be hundreds, maybe thousands. Based on the stories that he heard of the ants eating young animals and devouring birds as they emerged from their shells, Ziebach reasoned that these hundreds of millions—even billions—of ants could account for the plummeting quail population.

Ziebach broke the story on May 2, 1948 with an article in the news section of the *Mobile Press-Register* and a report in his usual "Out O'Doors" column in the sports section. "It hit like a thunderclap!" he wrote. "Where are our wits? What do we see when we look? Yes, we've previously seen the thousands of anthills—but it was just another colony of ants! It didn't register that what we saw was a general condition throughout the Coast. Along the roads, the mounds stand out like lights! And off the highway concealed by grasses, shrubs and trees are thousands upon thousands of ant hills containing billions of sharp-jawed insects armed with a sting which inflicts excruciating pain." He warned, "They are constant marauders which clean a small carcass in a few hours. They feed upon animals, birds, eggs, vegetation, seeds, crops—anything offering food. We can forget the idea of ever having rabbits, squirrels, quail or other ground-nesting birds—or tree nesters—until something is done to eradicate the horde of insects which has overrun the region."[88]

Ziebach's accounts tapped a nerve. ("Some of his articles," Creighton noted later, "could make your hair stand on end.")[89] Many southerners already worried that the bulldozer revolution was crushing their

---

87. Ibid.

88. B. Ziebach, "Is this the Answer to Disappearing Game?" *Mobile Press-Register,* 2 May 1948, 11B.

89. W. S. Creighton to W. F. Buren, 19 January 1969, Creighton papers.

favorite quarry, and now here was another threat. Responding to the desperation of Ziebach's warnings, E. O. Wilson began his study of the ant as the Alabama Department of Conservation contemplated ways to stop the insect.[90] Charged with protecting the state's game animals, the department worried that the ant would destroy what was left of the quail population, and was prompted to fund the work of Wilson and Eads.[91] Ziebach also contacted entomologist Clay Lyle and learned that Mississippi had successfully used chlordane to control the insect. Finding the USDA uninterested in studying the ant, Ziebach announced that he would lead Operation Ant against the bug, using chlordane to kill the ant wherever he found it.[92] Chemical insecticides, he said, were "the only straw that we could clutch to at the present time" and he hoped that they could be dropped from airplanes to blanket the infested areas.[93] The Department of Conservation joined his campaign and provided free chlordane to anyone who wanted it.[94]

Operation Ant and the portrait of the imported fire ant sketched in Ziebach's columns synthesized and extended ideas introduced by Wheeler (although there is no record that Ziebach ever read Wheeler) and the image of the fire ant offered by earlier entomologists. "Quietly it enters an area," Ziebach wrote of the insect, borrowing language used to describe communist invasions, "unnoticed or ignored until the entire countryside is blanketed with millions of mounds of a most vicious insect that knows no law except that of survival. They never rest or sleep, foraging the countryside unbelievable distances from the parent mound for anything that offers food."[95] It was "more damaging in this area than the boll weevil ever was or could be in the cotton-growing regions."[96] Evoking the title of military missions, Operation

---

90. Early fire ant notes, Wilson papers; B. Ziebach, "Department of Conservation Enters Tests Seeking Solution to Fire Ant Problem," *Mobile Press-Register,* 20 June 1948, 14B; "State Joins Fight against 'Fire Ants,'" *Mobile Press-Register,* 22 June 1948, 7A.

91. On the department's concerns, see B. Ziebach, "Department of Conservation Enters Tests Seeking Solution to Fire Ant Problem," *Mobile Press-Register,* 20 June 1948, 14B.

92. On the USDA's disinterest, see B. Ziebach, "Department of Agriculture Unable to Help in Combating Fire-Ant Infestation Here," *Mobile Press-Register,* 20 June 1948, 12B. On Operation Ant, see B. Ziebach, "Farmer Describes Infestation at Tanner-Williams," *Mobile Press-Register,* 20 June 1948, 12B.

93. Meeting of the advisory board, 17 February 1949, SG 17017, ALDoC papers.

94. Report of the advisory board of conservation, 13–15 August 1948, SG 17017, and Department of Conservation, Game, Fish, and Seafood Progress Report, 1 October 1949–1 February 1950, SG 17077, both in ALDoC papers.

95. B. Ziebach, "Monsters," *Mobile Press-Register,* 9 January 1949, clipping in early fire ant records file, Wilson papers.

96. B. Ziebach, "Department of Agriculture Unable to Help in Combating Fire-Ant Infestation Here," *Mobile Press-Register,* 20 June 1948, 12B.

Ant seemed an appropriate response. The ant was a killer and it needed to be destroyed.

In the years between Ziebach's first article on the insect and the initiation of the eradication campaign, a number of other newspapers and magazines published accounts of the ant. For the most part, these reflected not the qualified and confused reports of the entomologists who had studied the ant in the 1940s, but Ziebach's writings and the views and opinions of Wheeler and Wilson. One editorial cartoon in a Georgia newspaper, for example, portrayed the ant as an invading army, not unlike Union forces under General Sherman marching on Atlanta—another invasion of the South, another instance of total war (fig. 6). Alabama newspapers called the insect a "dread land scourge" that raised "devilish brood" and had staked "a hard-to-challenge lease on land."[97] Other press outlets labeled the ant "the red peril" and "fifth columnists," and worried that "[t]his ferocious little ant . . . has carried communism to the ultimate, and its actions suggest a certain cold-blooded intelligence."[98] The magazine *Senior Scholastic* proclaimed, "With the end of World War I in 1918, an undetected enemy agent stowed away on a South American freighter, and jumped ship in Mobile, Alabama. He devoted his first years to familiarizing himself with his new environment—so his subversive activities started slowly. But in recent years his sabotage has caused millions of dollars worth of damage."[99] The ant was now seen as a monster in the contemporary sense—not an omen, but a despicable creature inimical to human civilization.

From today's perspective, this image of the ant seems easy to dismiss, the frothy excess of an earlier, simpler era. But the portrait of the fire ant had roots that burrowed deeply into American thought and touched something vital—as evidenced by the 1954 movie *Them!*[100] In the film, nuclear tests mutated common desert ants into eight-foot-long giants that mindlessly killed—not unlike the mutated imported fire ants that besieged the South. The celluloid ants then spread, literally undermining the nation by taking up residence in the sewers of Los Angeles. One character worried, "We may be witnesses to a biblical

97. "Hitchhiker from South America—Alabama Farmers Face Worst Plague in History: Fire Ants," *Birmingham News*, 10 February 1957 (first two quotation); "Hordes of Fire Ants to Start Flying Soon," *Mobile Press-Register*, 18 February 1957 (third quotation), both in fire ant files (1957), PPC papers.

98. "Fire Ants Sting Congress into Action," *Dayton Daily News*, fire ant files (1957), PPC papers (first two quotations); J. Foster, "Secrets of the Fire Ant," *Mississippi Game and Fish* 19 (July 1957): 4 (third quotation).

99. "Insect Saboteur on the March," *Senior Scholastic* 76 (4 May 1960): 40.

100. "Them!" Warner Brothers, 1954.

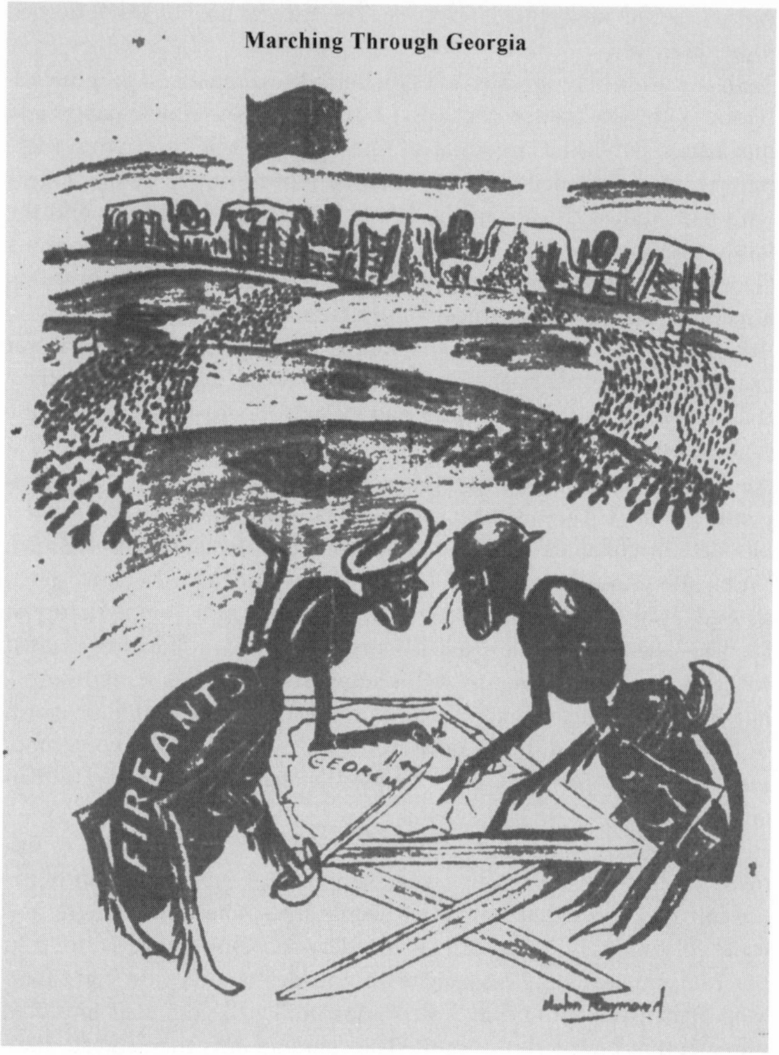
Marching Through Georgia
FIREANT
GEORGIA

Figure 6. This editorial cartoon from the *Macon (Ga.) News and Telegraph* made clear that the imported fire ant was an evil, invading force, evoking comparisons both with Sherman's March on Atlanta and contemporary fears of communist attack. As a military force, the ant needed to be met in martial terms. (From fire ant-news items file [1957], Plant Pest Control Division papers.)

prophecy come true: 'And there shall be destruction and darkness come upon creation, and the beasts shall reign over the earth.'" Only a martial response saved America, a combination of covert government operatives and military forces cornering the insects and destroying them in a fiery confrontation, not unlike the government and military operations needed to check the spread of communism. *Them!* was Warner Brothers' top-grossing film for the year and opened to critical applause for its realism.[101] *Newsweek* commented, "For many, if not most, devotees of science fiction, the great thing is not to make the fantastic seem superfantastic but to give it the strong illusion of reality. Judged by that standard, this is a right little fright of a picture," while *The New York Times* deemed the film "tense, absorbing and, surprisingly enough, somewhat convincing."[102] Belief that insects were antagonistic to human civilization, that mutant ants represented a dire threat, and that military action was a proper response to insect invasions braced an obviously fanciful film with a dose of verisimilitude.

The USDA and the Eradication Ideal

Bill Ziebach left little doubt that the imported fire ant was despicable, "a monster—a cancer—that is slowly and surely eating into the heart of the South"—deserving no quarter.[103] It was a powerful image: when an aide to Mississippi congressman Jamie Whitten visited some of the afflicted areas of the South in 1962, he expected to find not the quarter-inch-long insect that *Solenopsis invicta* was, but "quite a mammoth ant"—evidence of how monstrous the ant figured in the mind of the public.[104] But other descriptions of the ant, particularly by the USDA's entomologists, were more ambiguous, which prompts the question why the department chose to eradicate the ant. How did it evaluate the evidence and what calculus did it use to make its decision? Some have argued that the chemical industry convinced the USDA that it needed to eradicate the pest, not out of a sense of service to the public, but because it stood to make a handsome profit.[105] I disagree. I have been through thousands of boxes of USDA files in the National Archives, and I found very little material connecting the chemical industry to the campaign. Rather, I argue, to understand the development of the USDA's program requires looking at changes in the department taking place in the 1950s, as the ant spread.

101. B. Warren, *Keep Watching the Skies: American Science Fiction Movies of the Fifties*, vol. 1 (London: McFarland, 1982), 193–95.

102. *"Them!" Newsweek* 43 (7 June 1954): 56; J. R. Parrish and M. R. Potts, *The Great Science Fiction Pictures* (Metuchen, N.J.: Scarecrow Press, 1977), 318.

103. B. Ziebach, "Monsters," *Mobile Press-Register*, 9 January 1949, clipping in early fire ant records file, Wilson papers.

104. C. C. Fancher to E. D. Burgess, 27 November 1962, fire ant file, box 128, PPC papers.

105. For example, R. Carson, *Silent Spring* (Boston: Houghton Mifflin, 1962), 162.

During the 1950s, the ant's range extended from a few counties in Alabama, Mississippi, and Florida to encompass 20 million acres. As the ants reached Georgia, Texas, Arkansas, and Tennessee, farmers complained to county agents and politicians. The griping eventually reached the USDA.[106] Louisiana seemed especially hard hit. (Perhaps the ecological similarities between the swampy Bayou State and the South American interior aided the fire ant in Louisiana. Or, perhaps, the insect had established a beachhead in New Orleans unknown to entomologists, survived battles with the Argentine ant, and followed the bulldozer across the state from a second invasion site.) As the complaints piled up, officials in the USDA became increasingly anxious to revisit the ant problem.[107]

As the advance of the red imported fire ant prompted renewed concern, the USDA's entomological program underwent a reorganization that shaped the agency's response to the insect. In the early 1950s, agricultural officials divided the Bureau of Entomology and Plant Quarantine into the Entomological Research Division (ERD) and Plant Pest Control Division (PPC).[108] Headed by Edward Fred Knipling—who had overseen the USDA's first investigation of the imported fire ant—the ERD continued the tradition of those who had studied the insect in the late 1940s and early 1950s, approaching the control of insect pests as a biological problem, not a chemical one. Knipling's attempt to eradicate the screwworm fly, a terrible pest of southern cattle, illustrated the preferred approach. Studying the insect's life history, Knipling learned that the fly only bred once in its lifetime, a quirk of natural history that he exploited. Knipling used radiation to create millions of sterile male flies, reasoning that if he released enough into the wild, only infertile males would mate, the fly's reproductive output would grind to a halt, and the pest would go extinct. He could affirmatively answer Lyle's question of 1946 (whether it was time to pursue "the complete extermination" of some pests), but use the tools of a biologist, not a chemist.[109] Knipling and the ERD preferred a fire ant program that thoroughly investigated

106. For the complaints, see the correspondence in fire ant file, box 1; meetings 5 file, box 5; and fire ant file, box 15, all in PPC papers; insects file, box 2928, General Correspondence, 1906–1975, PI 191 and 1001 A-E (UD), Secretary of Agriculture papers, RG 16, National Archives and Records Administration, College Park, Md. (hereafter SOA papers).

107. The USDA's growing interest can be tracked in B. T. Shaw to A. J. Ellender, 14 May 1954; W. L. Popham to B. P. Livingston, 2 June 1954; E. D. Burgess to L. B. Johnson, 16 July 1954, all in fire ant file, box 1; L. F. Curl to E. E. Landry, 25 January 1955, fire ant file, box 15, all in PPC papers; "What Shall We Do about Ants?" *Alabama Journal*, 9 February 1957, 4.

108. E. D. Burgess, "Reorganization of the Plant Pest Control Branch," 4 October 1956, meetings, talks, papers file, box 48, PPC papers.

109. On Knipling, see J. Perkins, "Edward Fred Knipling's Sterile Male Technique for Control of the Screwworm Fly," *Environmental Review* 5 (1978): 19–37.

the insect's biology and that explored biological, cultural, and chemical control methods.[110]

The Plant Pest Control Division proposed a different method for dealing with the invasive ant. The PPC also embraced Lyle's call to eradicate insects, but, unlike Knipling, PPC entomologists followed Lyle in relying solely upon the new chemicals. E. D. Burgess, head of the PPC, announced in 1956, "Not only in the field of management are we experiencing the 'new look' but in many of our program activities, as well, where a noticeable change of emphasis has taken place. Wherever possible the old word 'eradication' has been taken from the shelf, dusted off, and placed in a position of prominence as a goal. This has not happened overnight but has been a gradual evolution linked hand in hand with the many new chemicals and improved practices that seem to make such an objective possible."[111] Burgess and his colleagues echoed Ziebach and Fortune, calling for a chemical attack to eradicate the imported fire ant.

Both the ERD and the PPC belonged to the USDA's Agricultural Research Service (ARS), but the PPC became more powerful than its sister agency, building what political scientists call an "iron triangle"—an alliance between government agencies, congressional committees, and lobbying groups—that nurtured and protected its programs.[112] Waldo Lee Popham was one point of the triangle: the PPC's first head and a staunch defender of the eradication ideal, Popham became deputy administrator of the ARS, where he supported the division's ambitions.[113] Jamie Whitten, a Mississippi Democrat who chaired the committee that controlled the USDA's budget, was another of the triangle's points. He doggedly defended funding for eradication programs.[114] Finally, state plant boards—agencies dedicated to protecting crops—agricultural commissioners, and the American Farm Bureau formed the

110. E. F. Knipling to E. D. Burgess, 9 September 1957, fire ant file, box 65, Entomology Research Division: General Correspondence, 1954–1958, 5 (UD), ARS papers.

111. E. D. Burgess, "Recent Developments in the Plant Pest Control Branch," 19–20 November 1955, meetings, talks, papers file, box 48, PPC papers. See also "USDA Scientist Calls Eradication the Best Way to Deal with Agricultural Pests," 27 December 1957, press releases file, box 63, Entomology Research Division: General Correspondence, 1954–1958, 5 (UD), ARS papers.

112. On the ARS, see E. G. Moore, *The Agricultural Research Service* (New York: Frederick A. Praeger, 1967). For iron triangles, see A. G. McConnell, *Private Power and American Democracy* (New York: Knopf, 1966); E. S. Redford, *Democracy in the Administrative State* (New York: Oxford University Press, 1969).

113. On Popham's views, see W. L. Popham and D. G. Hall, "Insect Eradication Programs," *Annual Review of Entomology* 3 (1958): 335–54.

114. C. J. Bosso, *Pesticides and Politics: The Life Cycle of a Public Issue* (Pittsburgh: University of Pittsburgh Press, 1987), 61–108.

third point of the iron triangle, lobbying Congress to support the PPC's programs.[115] With this support, the PPC entomologists dominated the ARS, absorbing resources as many entomologists in the ERD struggled for space and money.[116] The PPC initiated a number of eradication programs, targeting the Gypsy moth, the Japanese beetle, and Dutch elm disease, among others.[117] Just as American entomologists had helped to vanquish fascism during World War II, so would they now beat back another foe.[118] In 1957, the PPC proposed to eradicate the imported fire ant from all of the 20 million acres that it infested (fig. 7).[119]

The PPC's decision to eradicate the ant did not rely on any new information about the insect, or even any new information about the chemical insecticides that it proposed to use. After the Spring Hill, Alabama laboratory closed in 1953, the USDA undertook no more research on the insect.[120] Wilson and Ead's report proved that the ant was a pest, the agency said.[121] No more research was needed.[122] William Brown, Wilson's colleague at Harvard, disputed the adequacy of the early studies and requested $700 so that he and Wilson could travel south and investigate the situation.[123] "What constitutes adequate research is always open to discussion," one USDA official told Brown, but the early studies seemed convincing enough. The agency refused to

115. On the importance of state plant boards, see E. D. Burgess, "Reorganization of the Plant Pest Control Branch," 4 October 1956, meetings, talks, papers file, box 48; R. P. Colmer to W. L. Popham, 18 March 1957, fire ant files (1957), both in PPC papers.

116. For some of the ERD's difficulties, see C. W. Sabrosky, "Taxonomic Entomology in the U.S. Department of Agriculture," *Bulletin of the Entomological Society of America* 10 (1964): 211–220; "Report of the Ad Hoc Entomology Advisory Committee for the Smithsonian Institution," 23 January 1967, U.S. National Museum file, Philip Jackson Darlington papers, HUG (FP) 75.10, Harvard University Archives.

117. For an overview of these various eradication programs, see D. L. Dahlstein and R. Garcia, eds., *Eradication of Exotic Pests* (New Haven: Yale University, 1989).

118. For more on the link between World War II and the postwar eradication programs, see E. P. Russell III, "'Speaking of Annihilation': Mobilizing for War against Human and Insect Enemies, 1914–1945," *Journal of American History* 82 (1996): 1505–29.

119. "Report to Congress on the Fire Ant," 7 August 1957, regulatory crops-fire ant file, box 65, Entomology Research Division: General Correspondence, 1954–1958, 5 (UD), ARS papers.

120. W. C. Mcduffie to W. L. Brown, n.d., (*ca.* March 1957), fire ant file, box 65, Entomology Research Division: General Correspondence, 1954–1958, 5 (UD), ARS papers; E. D. Burgess to W. G. Eden, 16 April 1957, fire ant files (1957), PPC papers.

121. M. R. Clarkson to W. L. Brown, 30 January 1958, complaints-2-fire ant spray program file, box 195, PPC papers.

122. W. L. Popham, "The Imported Fire Ant," 5 August 1957, regulatory crops-fire ant file, box 752, General Correspondence, 1 (UD), ARS papers.

123. W. L. Brown to M. R. Smith, 21 March 1957, fire ant file, box 65, Entomology Research Division: General Correspondence, 1954–1958, 5 (UD), ARS papers; W. L. Brown and H. Levi to M. R. Clarkson, 7 January 1958, complaints-2-fire ant spray program file, box 195, PPC papers.

Figure 7. **The relationship between humans and insects, as a number of historians have noted, was cast as a military battle.** In the years after World War II, humans seemed to have gained the upper hand with the invention of DDT and other new chemical insecticides. The caption reads, "Modern Warfare Is Really Rough."(From the *Arkansas Gazette*, 29 July 1957, 4a. Used by permission of the *Arkansas Gazette*.)

give Brown and Wilson the grant.[124] E. F. Knipling also sought to initiate a research program, asking the PPC for $54,500 to study the insect's biology "so as to take advantage of unknown weak links that may be useful in insecticide or other control procedures."[125] Burgess dismissed

124. M. R. Clarkson to W. L. Brown, 30 January 1958, complaints-2-fire ant spray program file, box 195, PPC papers.

125. E. F. Knipling to T. C. Byerly, 6 May 1957 and 10 September 1957; E. F. Knipling to E. D. Burgess, 9 September 1957 (quotation), all in fire ant file, box 65, Entomology Research Division: General Correspondence, 1954–1958, 5 (UD), ARS papers.

his ideas, however, saying "basic research" was of "no immediate value."[126] Chemical insecticides were enough. The PPC's only concession was to hire some ERD entomologists to study the insecticides used against the ant in order to find the cheapest, most potent chemical formulation.

The ERD entomologists, though, would not start their work in earnest until after the eradication program had begun; the PPC entomologists were convinced that the insecticides it proposed to use were safe enough—and the ant dangerous enough—to justify beginning the program without preliminary research. "Sufficient research has been done on which to base an eradication program," said Waldo Lee Popham in August 1957, but not everyone agreed.[127] The Southern Plant Board's Ross Hutchins cautioned, "As far as I know, no research is available to determine just what would be the effect of widespread application of either dieldrin or heptachlor. It is my guess that many quail, at least, would be killed."[128] Early in 1958, Knipling offered a more stern warning: heptachlor used to control grasshoppers on the western range appeared in meat and milk at levels higher than allowed by law, even though the chemicals were used at one-sixteenth the dose employed in the Southeast.[129] Burgess shrugged off Knipling's opinion: research would follow, but it was more important to start the program.[130] For the PPC's entomologists, a few wildlife deaths were worth the risk. As one eradication proponent wrote, "To clothe and feed [the American] population, man must maintain his position of dominance, and our agricultural production must continue to increase even at the expense of the further displacement of native plants and animals."[131]

126. E. D. Burgess to E. F. Knipling, 7 October 1957, fire ant files (1957), PPC papers.

127. W. L. Popham, "The Imported Fire Ant," 5 August 1957, regulatory crops-fire ant file, box 752, General Correspondence, 1 (UD), ARS papers.

128. R. E. Hutchins to Southern Plant Board, 12 November 1957, fire ant files (1957), PPC papers.

129. On this research, see Entomology Research Division, "Residues in Fatty Tissues, Brain, and Milk of Cattle from Insecticides Applied for Grasshopper Control on Rangeland," *Journal of Economic Entomology* 52 (1959): 1206–10.

130. On the ERD's protests, see C. H. Hoffman to E. F. Knipling, September; E. F. Knipling to E. D. Burgess, 9 September 1957; E. F. Knipling to E. D. Burgess, 22 October 1957; A. W. Lindquist to C. N. Smith, 22 October 1957; C. N. Smith to A. W. Lindquist, 24 October 1957, all in fire ant file, box 65, Entomology Research Division: General Correspondence, 1954–1958, 5 (UD), ARS papers; E. F. Knipling to E. D. Burgess, 2 December 1957, fire ant files (1957), PPC papers. On the PPC's response, see G. J. Haeussler, office memo, 23 August 1957; E. D. Burgess to E. F. Knipling, 7 October 1957; and E. D. Burgess to E. F. Knipling, 15 November 1957, all in fire ant files (1957), PPC papers; and Knipling to Burgess, 22 October 1957, and C. N. Smith to A. W. Lindquist, 30 October 1957, both in fire ant file, box 65, Entomology Research Division: General Correspondence, 1954–1958, 5 (UD), ARS papers.

131. G. C. Decker, "Pesticides Relationship to Conservation Programs," *National Agricultural Chemicals Association News and Pesticide Review*, May 1959: 7.

Knipling, though, would not be quieted: no one had ever tried to eradicate an insect from 20 million acres and there was no way to gauge the
effects of the insecticide on domestic animals and wildlife. If things
went badly, he said, and someone sued, the USDA would lose the
case.[132]

With no new information about the ant or the insecticides used to
kill it, the question remains: Why did the PPC decide to eradicate the
pest? It seems that the decision to launch such an audacious program
derived mostly from political and bureaucratic concerns. Although the
PPC had gained substantial power within the USDA, eradication advocates had established themselves in colleges, universities, and state governments, and chemical insecticides had been used successfully during
World War II, the eradication ideal had not yet proven its worth. Only
a few small insect infestations had been eradicated, and some USDA
campaigns had attracted criticism. In New York a band of citizens sued
the USDA, claiming that the Gypsy moth eradication program threatened human health.[133] The eradication ideal had been especially inept
in the South. A few years before drawing up plans to eradicate the
imported fire ant, the USDA had tried to eradicate the white-fringed
beetle, another South American import that ravaged the South. The
department had failed to excite popular interest in the program and
the beetle continued to spread.[134] Eradicating the ant, I think, was a
way of redressing this failure and justifying the PPC's embrace of both
chemical insecticides and the eradication ideal. C. C. Fancher, who
headed the white-fringed beetle program when it collapsed, and was
then chosen to head the fire ant program, considered the attack on
Solenopsis an "opening wedge" for renewing the eradication of the
white-fringed beetle: by carpeting the South with insecticides, the PPC
entomologists would kill the beetle *and* the ant—affirming the wisdom
of the eradication ideal and the importance of the Plant Pest Control
Division.[135] Much depended on the eradication campaign's success.

132. C. H. Hoffman, memo, 30 April 1958, insects affecting man and animals branch fire
ant file, box 84, Entomology Research Division: General Correspondence, 1954–1958, 5 (UD),
ARS papers.

133. On the Gypsy moth trial, see "Gypsy Moth Case," *Journal of Agricultural and Food Chemistry* 6 (1958): 496–98; T. R. Dunlap, *DDT: Scientists, Citizens, and Public Policy* (Princeton: Princeton University Press, 1981), 87–89; C. J. Bosso, *Pesticides and Politics: The Life Cycle of a Public
Issue* (Pittsburgh: University of Pittsburgh Press, 1987), 81–83.

134. On problems with the white-fringed beetle eradication program, see the material in
white-fringed beetle file, box 5; white-fringed beetle file and white-fringed beetle (correspondence with J. Lloyd Abbot) file, both in box 29, all in PPC papers.

135. The quotation is from L. F. Curl, memo, fire ant files (1957), PPC papers. Belief that the
fire ant eradication program would help control the white-fringed beetle was widespread. For

Fancher wrote, "It would be a disgrace to entomologists of this country to permit the imported fire ant to become established."[136] In particular, PPC entomologists would be disgraced, and the need for their bureaucracy and their style of entomology called into question.

The decision to eradicate the fire ants, though, was not all selfish. While it was an attempt to redress a policy failure and substantiate the claims of a select group of bureaucrats, it also aimed to help southerners. And, in any event, killing bugs was deemed patriotic, as historian Edmund Russell has made clear in his study of the links between entomology and war during the middle of the twentieth century.[137] It was an article of faith at the USDA that American democracy grew from the soil: agriculture was a Cold War weapon. Byron Shaw, head of the ARS, for example, said in 1958, "I think the times were described rather aptly a year or so ago, when a Soviet premier told an American television audience that communism would win its contest with capitalism when the Soviet's per-capita production of meat, milk and butter surpassed that of the United States. He was really saying that a nation is as strong as its agriculture."[138] Insects sabotaged America's ability to compete in global politics. One member of the House of Representatives, for example, called insects an "evil force" that reduced American agricultural abundance, lowering the standard of living and making people dissatisfied. "I do not need to tell you that dissatisfaction breeds communism," he warned.[139] The threats posed by the imported fire ant fit easily into this ideology since it had repeatedly been linked to communism. Louisiana representative Theo A. Thompson, for instance, told Whitten's committee, "We, in the United States, have been prone to spend billions of dollars in the matter of fighting communism and the results of the atomic bombs and radiation and all, but should we

other examples, see R. A. Roberts to E. L. Ayers, 24 August 1954, fire ant file, box 1; C. C. Fancher to E. D. Burgess, 25 February 1957, fire ant files (1957); M. Brunson to C. W. Telfair, 26 February 1957, white-fringed beetle-southern region file, box 81; Charles Lincoln to John W. White, 5 April 1957, fire ant files (1957); J. W. Patterson to C. C. Fancher, 12 April 1957, white-fringed beetle-southern region file, box 81; C. C. Fancher to E. D. Burgess, 3 May 1957, fire ant-southern region file, box 197; C. C. Fancher to Area Supervisors, 30 April 1958, fire ant file, box 197; C. C. Fancher to J. F. Spears, 7 June 1960, fire ant files, box 225, all in PPC papers.

136. C. C. Fancher to E. D. Burgess, 20 May 1958, fire ant file, box 197, PPC papers.

137. E. P. Russell III, *War and Nature: Fighting Humans and Insects with Chemicals from World War I to Silent Spring* (Cambridge: Cambridge University Press, 2001).

138. B. T. Shaw, "Development of Research Facilities and the Role of Regional Laboratories in State and Federal Programs," 10 November 1958, speeches file, box 34, Entomology Research Division: General Correspondence, 1954–1958, 5 (UD), ARS papers.

139. Quotation from C. J. Bosso, *Pesticides and Politics: The Life Cycle of a Public Issue* (Pittsburgh: University of Pittsburgh Press, 1987), 69.

not spend some of our money to protect our country against these parasites. Would it not be funny to spend all these billions of dollars fighting communism and building the atomic bombs and then be eaten up by the Argentine ant?"[140] Thus, Fancher could say without irony that the eradication of the imported fire ant would protect "the American way of life."[141] Deciding to eradicate the ant tapped that same vital part of the American psyche that made *Them!* seem realistic: the ant was the enemy (fig. 8).

Starting the Program

A shared sentiment about the ant, insecticides, and the eradication ideal cemented relations among the PPC, Congress, and the public. But, while congruent opinions were necessary for the eradication program to work, they were not alone sufficient to initiate the campaign. Other obstacles still stood between the PPC and the eradication of the ant, most notably Ezra Taft Benson, the secretary of agriculture, and public apathy. Benson supported the use of insecticides—he called them "marvels" and mused about the day that chemical companies would sow the clouds with them so that when it rained crops were sprayed, too—but he hated huge government programs.[142] He forbade agricultural officials from asking Whitten to fund the fire ant eradication program.[143] The PPC, then, needed specific legislation from Congress authorizing the program and a steady stream of money to ensure that the program could continue despite Benson's objections. Even with a dedicated flow of funds, though, the program could still falter if it did not have universal support. Eradication, as agricultural officials often noted, necessitated that all land be sprayed, "without regard to location, land use, or ownership."[144] If any land was left untreated, some

140. U.S. Congress, Subcommittee on Research and Extension of the Committee on Agriculture, *Plant Pests, Control, and Eradication,* 85-1, 1957, 41.

141. C. C. Fancher to E. D. Burgess, 27 June 1958, fire ant-wildlife losses file, box 198, PPC papers.

142. For Benson's view of insecticides, see E. T. Benson, Address to the Joint Meeting of the Commercial Chemical Development Association and the National Agricultural Chemical Association, 21 November 1958, information-press releases file, box 82, Entomology Research Division: General Correspondence, 1954–1958, 5 (UD), ARS papers. For his political views, see E. T. Benson, *Freedom to Farm* (Garden City, N.Y.: Doubleday, 1960); E. T. Benson, *Cross Fire: The Eight Years with Eisenhower* (Garden City, N.Y.: Doubleday, 1962); E. L. Schapsmeier and F. H. Schapsmeier, *Ezra Taft Benson and the Politics of Agriculture: The Eisenhower Years, 1953–1961* (Danville, Ill.: Interstate Printers and Publishers, 1975).

143. G. J. Haeussler, memo, 23 August 1957, fire ant files (1957), PPC papers.

144. The phrase was ubiquitous, but for one example, see W. L. Popham to L. Hill, 18 January 1957, regulatory crops-1-fire ants file, box 752, General Correspondence, 1954–1966, 1 (UD), ARS papers.

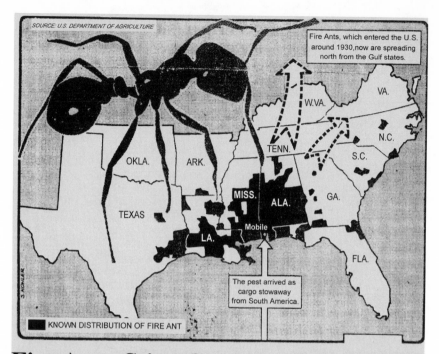

SOURCE: U.S. DEPARTMENT OF AGRICULTURE

Fire Ants, which entered the U.S. around 1930, now are spreading north from the Gulf states.

The pest arrived as cargo stowaway from South America.

■ KNOWN DISTRIBUTION OF FIRE ANT

OKLA. ARK. TENN. W.VA. VA. N.C. S.C. MISS. ALA. GA. TEXAS Mobile LA. FLA.

Fire Ants Sting Congress Into Action

BY ROGER GREENE

A SCOURGE of fierce-biting Argentine "fire ants," spreading north from the Gulf states, has prompted Congress to act with almost as much speed as if the lawmakers had been stung by the fiery red varmints.

Experts say the ants—whose sting has hospitalized grownups and caused the death of a three-year-old boy in New Orleans—may eventually reach as far north as Canada.

Spurred by outcries from Dixie, the Senate and House quickly passed legislation authorizing the Department of Agriculture to cooperate with states in combatting the plague.

The only hitch is that the lawmakers in their current penny-pinching drive failed to provide a red cent to fight the red peril.

There are indications, however, that the Budget bureau will soon ask Congress to appropriate funds enabling the Agriculture department to join the affected states in a multimillion-dollar campaign to stem the onslaught.

Latest reports show that the aptly named fire ants—its sting inflicts fiery pain, perhaps worse than a hornet sting, and raises angry welts—already has infested 27 million acres in nine states.

A voracious pest to crops, livestock and humans, the tiny flame-colored ant has sent its legions on devastating forays into Mississippi, Louisiana, Alabama, Georgia, eastern Texas and Florida.

* * *

TO A LESSER EXTENT, it has also been reported from scattered locations in Arkansas, South Carolina and North Carolina and the southwest tip of Tennessee.

First reported near Mobile, Ala., in 1930, the ant is generally believed to have come to this country from Argentina. For some years it was little more than a minor pest; then suddenly it exploded into a major infestation causing millions of dollars in damage annually.

By way of illustrating the rapid expansion of the threat, the Agriculture department says that up to 1949 the fire ant had infested only nine counties in Alabama but had

spread to 26 counties by 1953 and to 51 counties by 1957.

Rep. Roberts (D-Ala.) told a House agriculture subcommittee the ants have infested more than 13 million of Alabama's total 32½ million acres.

"This is a much more serious problem than the boll weevil," he said.

Rep. Elliott (D-Ala.) said it was estimated the fire ant had already caused more than 25 million dollars worth of damage in his state.

"It is really a terrible thing," Rep. Boykin (D-Ala.) told the subcommittee. "The fire ants have killed worlds of quail on our place, wild turkeys and also calves and young deer.

"If we don't watch out it is going to cover the whole United States."

Expert testimony supported Boykin's warning that the plague could become a nationwide problem.

* * *

THE FIRE ANT, he explained, builds its homes as deep as five feet underground or "far below the frost level" in most American farm areas.

So far, the pest has been

spotted as far north as Wake county, N. C.

Ranging in size from one-eighth to one-quarter inch long, the Argentine ants have won something of a reputation as "fifth columnists."

Unlike army ants and other insects that travel en masse invading a territory like marching soldiers, the fire ant usually infiltrates a new area unnoticed and gradually builds up to formidable proportions.

Entomologists say the ant probably arrived in this country as cargo stowaways from South America and spread out by flying and crawling, drifting downstream on logs and by traveling aboard cars, trucks, trains and planes.

The fire-ant "workers" don't fly, but the winged queen, after mating, takes off and promptly starts up a new colony elsewhere, thus spreading the affliction.

In homes the ants eat meat, butter, cheese and nuts.

No natural enemy has been found to destroy the Argentine invaders, but the Agriculture department says three poisons have been found effective against them. However, it costs as much as $4 an acre to banish the pest.

ants might survive and spread, mocking the eradication ideal. Levi Curl, the PPC official in Washington, D.C. directly responsible for the fire ant program, said, "There is one basic principle which would be essential" for eradication: "The State would have to be in a position to require that treatments would have to be made to every square foot of ground needing treatment. . . . The plan would have to be mandatory rather than voluntary in order to succeed."[145] Coordination with the state plant boards gave the USDA authority to spray land without the owner's permission, but the law mandated that the USDA could only pay for half of the program.[146] State and local communities had to pay the other half. Thus, even those without ant problems had to pay for the eradication campaign, and the USDA needed to convince them of the program's necessity to maintain funding.

As I see it, the PPC overcame Benson's opposition with some deft political maneuvering. In early 1957, Whitten's committee granted the USDA power to eradicate the imported fire ant, but did not appropriate any funds, although the American Farm Bureau had requested $10 million and the Southern Plant Board $12 million.[147] Whitten's committee assumed that the USDA could tap a $2 million emergency fund.[148] The PPC, though, lamely responded that the fire ant had been in the country for decades and so did not constitute an emergency: the emergency funds were off limits.[149] Once the eradication program began, PPC

145. L. F. Curl, "Setting up a Plant Pest Eradication Program," 6 October 1956, meetings, talks, papers file, box 48, PPC papers.

146. On the power vested in the states, see U.S. Congress, Subcommittee on Fisheries and Wildlife Conservation, Committee on Merchant Marine and Fisheries, *Coordination of Pesticide Programs*, 86-2, 1960, 71–72.

147. On the Southern Plant Board, see S. J. McRory to A. W. Todd, 24 April 1957, SG 8836, Department of Agriculture and Industry files, Alabama Department of Archives and History, Montgomery, Ala. (hereafter ADAI papers). On the Farm Bureau, see "$10 Million Sought to Fight Fire Ants," *Montgomery Advertiser*, 8 March 1957, fire ant news clippings file, box 65, Entomology Research Division: General Correspondence, 1954–1958, 5 (UD), ARS papers. Testimony in support of the eradication program can be found in U.S. Congress, Subcommittee on Research and Extension, Committee on Agriculture, *Plant Pests, Control and Eradication*, 85-1, 1957.

148. C. J. Bosso, *Pesticides and Politics: The Life Cycle of a Public Issue* (Pittsburgh: University of Pittsburgh Press, 1987), 87.

149. For the PPC's response, see W. L. Popham to L. Hill, 10 May 1957, fire ant files (1957); "Imported Fire Ant," memo, 7 June 1957, appropriations (budget) June-file, box 58, both

---

Figure 8. "Fire Ants Sting Congress into Action." The ant's invasion was perceived in militaristic terms. The arrows showing its movement echoes the cartography of World War II battles, and the monstrously large drawing of the insect heightens the sense that it imperiled the South. The USDA provided this figure to many southern papers as part of its attempt to solidify support for the eradication effort (From fire ant-news items file [1957], Plant Pest Control Division papers.)

officials were not above using creative financing to maintain the program's momentum, so it seems odd that they refused to stretch the definition of "emergency" to get the operation off the ground. The ant may not have been newly imported to the United States, but it was causing a problem that many characterized as an emergency.[150] The PPC's response, though, makes sense if one sees it as an attempt to circumvent Benson. It seems likely that the PPC refused to tap the emergency funds in hopes that angry southerners would demand that Whitten provide money for the eradication campaign. If that was the case, the ploy worked. The press lambasted the government for its failure to do something about the ant and southern legislators worked feverishly to fund the program.[151] In the end, Congress appropriated $2.4 million for the fire ant program, with more to come each year. The PPC had received authorization and money for the program without making a formal request.

To ensure a public mandate, the PPC instituted a publicity campaign. In early 1957, the USDA assigned a public relations specialist to coordinate press releases, while entomologists associated with the department helped to foster support.[152] The department ordered one thousand copies of one of Wilson's popular articles about the insect to distribute.[153] And USDA officials emphasized that the ant posed a threat not just to the South, but to the entire country. J. F. Spears, working in the PPC's southern regional office, warned that the fire ant could reach the Fortieth Parallel, "and it is possible the ant could become established north of that line," he said.[154] *Time* magazine responded with a

---

in PPC papers; "Report to the Congress on the Fire Ant," 2 August 1957, fire ant file, box 65, Entomology Research Division: General Correspondence, 1954–1958, 5 (UD); draft response, 15 August 1957, regulatory crops-fire ant file, box 752, General Correspondence, 1954–1966, 1 (UD), both in ARS papers.

150. For an admission that the PPC bent accounting rules, see C. C. Fancher to PPC Supervisors, 10 September 1962, fire ant files, box 128, PPC papers.

151. "Neglect in Fighting Fire Ants," *Arkansas Democrat,* 30 May 1957; "Fire Ant Program Snagged in Capital," *Commercial Appeal,* 28 May 1957, both in fire ant files (1957), PPC papers. For the work of the congressional legislators, see F. Boykin to A. W. Todd, 26 August 1957, SG 8836, ADAI papers.

152. On the assignment of the public relations officer, see "Recent Developments in the Plant Pest Control Division," 30 August 1957, Agriculture Secretary's Office file, box 58, PPC papers; for examples of the relations between county agents and locals, see C. C. Fancher to E. D. Burgess, 25 February 1957; R. L. Boyd to E. T. Benson, 5 September 1957, both in fire ant files (1957), PPC papers; P. Daniel, *Lost Revolutions: The South in the 1950s* (Chapel Hill: University of North Carolina Press, 2000), 80. See also J. C. Devlin, "U.S. Helping South in Fire Ant Fight," *New York Times,* 23 December 1957, 23.

153. J. S. Riss to J. Purcell, 11 June 1959, fire ant-8 file, box 213, PPC papers.

154. "War against Fire Ants about to Begin," *New Orleans Times-Picayune,* 27 October 1957, fire ant files (1957), PPC papers.

picture of the ant and a question: "Headed for Pennsylvania?"[155] (Dr. M. R. Clarkson, a deputy administrator with the USDA, testified to Congress, "We do not know what the farthest northward spread might be, but it is our belief that it would spread to any part of the United States where farming is conducted.")[156] Marion Smith tried to temper some of the rhetoric—predictions of the ant's spreading across the nation were unfounded speculation, he said—but PPC officials ignored him.[157] The ant was a danger, the eradication advocates said, and like the insects in the movie *Them!* or invading communists, it could only be stopped with martial force. A widely reprinted department press release noted, "Uncle Sam is ready to use a fleet of 60 planes to go to war against the dreaded fire ant. . . . Only the modern airplane, dropping insecticides on twenty million acres in the critical area, can hope to stop the menace."[158] The department's entomologists planned to establish a quarantine around the ant's perimeter to prevent it from escaping the South, then "pattern bomb" from the periphery inward, concluding with a mop-up operation to hunt down and kill any stragglers: this was, in essence, the fire ant wars.[159] If all went well, in twenty years and at a cost of $100 million, the USDA would have sprayed heptachlor onto 20 million acres and eradicated the pest.

The Imported Fire Ant on the Ground

The USDA's decision to eradicate the imported fire ant rested on an image of the insect stripped of ambiguity. The department did not mention its own reports from the 1940s that questioned the amount of damage attributable to the insect and the practicality of eradication. Instead, it relied mostly on the studies of E. O. Wilson, which, in turn, had been based mainly on the testimony of farmers. In later years, the myrmecologist himself said that he put "little faith" in these estimates.[160] But he did observe how the ant attacked crops, protected aphids, gathered seeds, and built large mounds that interfered with tractors. So, how bad was the ant? Was it really a threat to all

---

155. "Fiery Invader," *Time* 69 (18 March 1957): 84–85.

156. U.S. Congress, Subcommittee on Research and Extension of the Committee on Agriculture, *Plant Pests, Control, and Eradication*, 85-1, 1957, 23.

157. M. R. Smith to W. S. Creighton, 9 December 1957, Creighton papers.

158. "U.S. Ready to Join Aerial War against Fire Ant," 2 April 1957, fire ant files (1957), PPC papers.

159. This description of the program is from J. L. George, *The Program to Eradicate the Imported Fire Ant* (New York: Conservation Foundation, 1958), 15—16, and "Report to Congress on the Fire Ant," 2 August 1957, regulatory crops-fire ant file, box 65, Entomology Research Division: General Correspondence, 1954–1958, 5 (UD), ARS papers (quotation).

160. E. O. Wilson to P. Charam, 14 July 1964, fire ant file, box 247, PPC papers.

agriculture and ground-nesting birds, as Ziebach had said? Or was it merely a nuisance? Assessment is difficult, given the lack of rigorous surveys, the confusion between native and imported insects, and the frequency of exaggeration, but it is possible to make an informed guess.

That the ant was a pest, sometimes a terrible one, is incontrovertible. Wilson saw the damage himself, as did Fortune, Smith, and Ziebach. Reports came from across the South. On one Alabama farm, for example, the ant destroyed three acres of corn in 1954.[161] And in Mississippi, J. T. Brown shut down his sheep farm because the ant harassed his animals too much and prevented the use of his farm machinery—convincing evidence that the insect caused some problems, even if it was not the sole cause of Brown's difficulties.[162] Ecological theory predicts these consequences of the ant's invasion. Freed from parasites, predators, and diseases, presented with an inviting environment, its population exploded.[163] Opportunistic, it ate what food was available. And numerous, it inflicted multiple stings on people: in 1955, a New Orleans boy died twenty-nine hours after fire ants stung him; two years later, ant stings sent three soldiers from Maxwell Air Force base to the hospital.[164] Bees and wasps probably killed about the same number of people, but the ant was new, scary, and seemingly ubiquitous.[165] Calls for federal assistance were frequent and earnest. Farmers joined into clubs to fight the fire ant, volunteered to help the eradi-

161. C. C. Fancher to W. G. Bruce, 20 May 1954, fire ant file, box 1, PPC papers.

162. J. T. Brown to J. W. Patterson, 15 November 1957, fire ant files (1957), PPC papers.

163. The classic source on the behavior of introduced species is Charles Sutherland Elton, *The Ecology of Invasions by Animals and Plants* (London: Chapman and Hall, 1977 [1958]). See also H. Mooney and J. Drake, eds., *Ecology of Biological Invasions of North America and Hawaii* (New York: Springer-Verlag, 1986); J. Carey et al., eds., "Special Issue: Invasion Biology," *Conservation Biology* 78 (1996).

164. "Fire Ants Spread as Pastures Shrink," *Montgomery Advertiser,* 17 February 1957; "Maxwell Reports Plague of Fire Ants; 3 in Hospital," [newspaper unknown,] 14 March 1957, both in fire ant files (1957), PPC papers.

165. Data on mortality from insect stings is difficult to compile since the cause of death is often attributed to sequelae of the sting, rather than the sting itself. See A. I. Neugut, A. T. Ghatak, and R. L. Miller, "Anaphylaxis in the United States: An Investigation into Its Epidemiology," *Archives of Internal Medicine* 161 (2001): 2046–47. According to this same source, bees, wasps, and ants kill from forty to one hundred people each year; a 1989 survey showed that fire ants killed thirty-two people. See R. B. Rhoades, C. T. Stafford, and F. K. James Jr., "Survey of Fatal Anaphylactic Reactions to Imported Fire Ant Stings. Report of the Fire Ant Subcommittee of the American Academy of Allergy and Immunology," *Journal of Allergy and Clinical Immunology* 84 (1989): 159–62. If this is number is representative, then the imported fire ant, even occupying only part of the United States, accounts for a goodly proportion of the deaths caused by insect stings. But, there is at least one caveat: studies of fire ant stings are much more intense than studies of bees and wasps.

cation campaign, and invented their own means of dealing with the insect.[166] At a meeting of the Southern Plant Board in early 1957, a half-hour was set aside for a consideration of the ant: the discussion continued for two hours, an indication of how important the ant had become.[167]

The ant especially bothered cattlemen, probably because it thrived on open pastures.[168] A farm journalist reported, "Cattlemen with their widespread pastures have suffered most from the onslaught."[169] In 1962, the USDA confirmed that the ant killed a cow—the first documented case—and in earlier years there were plenty of less spectacular but more irritating encounters between insect and rancher.[170] Mowing was difficult on fields studded with dozens of mounds per acre.[171] In 1957, the president of the Alabama Cattleman's Association said that the ant rendered about 25 percent of pastures unusable.[172] Even if this claim stretched the truth some, it indicates that *Solenopsis invicta* caused significant problems.

Although it is clear that the fire ant was a real threat, there were also obvious exaggerations of the threat it posed even before the USDA started its publicity campaign. In 1955, for example, a Louisiana farmer worried that his farm was "literally undermined with ants, which people say will eat anything I try to raise."[173] Fears such as these propagated over the years, fed in part by those such as Air Force captain Warren Papin, who confused the imported fire ant with the army ant. In 1957, Papin told a newspaper that in South America *Solenopsis* "will sometimes attack large animals and kill them with their multiple bites. Apparently, they are meat eaters since the carcasses were stripped."[174]

166. On these methods, see the correspondence in fire ant file, SG 8836, ADAI papers.

167. F. S. Arant, Memo to E. V. Smith and J. L. Lawson, 12 March 1957, box 2, Ralph B. Draughon papers, RG 107, Auburn University Archives.

168. On the experience of cattlemen, see "Elements of Myth on the Fire Ant Peril," *Montgomery Advertiser,* 11 March 1957; "Congressional Fire Ant Aid Requested," *Montgomery Advertiser,* 23 March 1957, both in fire ant news clippings file, box 65, Entomology Research Division: General Correspondence, 1954–1958, 5 (UD), ARS papers; Memo from Representative Selden's Office, 23 March 1957, fire ant file, SG 8836, ADAI papers.

169. A. Rankin, "The Great Fire-Ant Invasion," *Reader's Digest* 71 (September 1957): 76.

170. On the 1962 discovery, see W. G. Cowperthwaite to E. D. Burgess, 30 November 1962, fire ant file, box 128, PPC papers.

171. "Special Report of Tom Lemon," n.d., fire ant files (1957), PPC papers.

172. "Congressional Fire Ant Aid Requested," *Montgomery Advertiser,* 23 March 1957, fire ant news clippings file, box 65, Entomology Research Division: General Correspondence, 1954–1958, 5 (UD), ARS papers.

173. W. E. Clement to Brooks, 11 January 1955, fire ant file, box 15, PPC papers.

174. "Maxwell Reports Plague of Fire Ants; 3 in Hospital," [newspaper unknown,] 14 March 1957, fire ant files (1957), PPC papers.

This report was patently false, but others absorbed the fear it communicated. A Texas rancher, for instance, heard "that there are some areas in the Eastern states so badly infested that should a cow have a calf, before the calf can get up and move the ants will kill it"; another cattleman claimed that the ant had killed two, perhaps three of his calves and left only bones.[175] The stories are unlikely, and their inexactness identifies them as rumors, but they do highlight the widespread fear of the insect. Fueled by these tales, southern newspapers printed scores of articles detailing the horrors perpetrated by the "imported devil insect."[176] "Alabama faces worst plague in history," proclaimed one, while another reported that the ant had "become farmers' number one pest," causing $25 million worth of damage.[177] Frustrated, the Alabama rancher and farm bureau member Julian Elgin told a newspaper, "The serious threat to the economy of our area justifies any act which we may take to rid our country of this terrible pest."[178]

A rebuke to this hyperbole can be found in the responses of many southerners. While some undoubtedly suffered, losing land and crops, others shrugged at the invasion. In the early and mid-1950s, for example, Louisiana subsidized the sale of chlordane (it cost farmers just fifteen cents per pound: thirty cents to treat an acre of cultivated land) but only a few farmers purchased the insecticide (fig. 9).[179] And in Butler County, Alabama, only fifteen out of sixteen hundred farmers— representing .0032 percent of the land area—wanted an eradication program; the other 1,585 farmers wanted just enough pesticide to control the ant on their land, but saw no reason to exterminate it completely.[180] A PPC official admitted that similar sentiment could be found throughout northern Alabama.[181] The ant may have been a problem, but it was not so troublesome that people were

175. N. W. Mitchell to E. T. Benson, 10 July 1958, fire ant file, box 22, Wildlife Research Correspondence, 1934–1966, 254 (UD); United States Fish and Wildlife Service papers, RG 22, National Archives and Records Administration, College Park, Md.

176. "Fire Ants Spread as Pastures Shrink," *Montgomery Advertiser,* 17 February 1957, fire ant files (1957), PPC papers.

177. "Hitchhiker from South America—Alabama Farmers Face Worst Plague in History: Fire Ants," *Birmingham News,* 10 February 1957; "Fire Ants Become Farmers' Number One Pest," *Alabama Farmer's Bulletin,* 15 March 1957, both in fire ant files (1957), PPC papers.

178. "$10 Million Sought to Fight Fire Ants," *Montgomery Advertiser,* 8 March 1957, fire ant news clippings file, box 65, Entomology Research Division: General Correspondence, 1954–1958, 5 (UD), ARS papers.

179. On the program in Louisiana, see A. J. Ellender to B. T. Shaw, 30 April 1954, fire ant file, box 1; E. S. Landry to E. E. Willis, 8 January 1955, fire ant file, box 15, both in PPC papers.

180. J. P. Henderson to A. W. Todd, 10 April 1958, fire ant file, box 198, PPC papers.

181. C. C. Fancher to E. D. Burgess, 2 May 1958, fire ant file, box 198, PPC papers.

Figure 9. Although there was broad support for the eradication of the imported fire ant, not everyone was happy about the program's cost. This editorial cartoon from a Louisiana newspaper expressed concern about the expense of the program: the state had ants in its pants, and the insects were going for the wallet. (From *New Orleans Times-Picayune*, 16 April 1957, fire ant-news item file [1957], Plant Pest Control Division papers.)

willing to pay any amount—or even in some cases, small amounts—to eradicate it.

The image that emerges, then, is of an ant that was a pest, sometimes a serious one, but one that did not cause problems uniformly across the South. In some places it could be a bane to farmers, but in others it provoked little concern. The early USDA investigations, although hurried, underfunded, and not repeated again until the late 1950s, after the eradication program had begun, were largely correct: the ant damaged crops, made farming difficult, and occasionally attacked or killed vertebrates, including humans; but complaints about the insect were frequently exaggerated. A survey of insect pests in the South substantiates this point. In May 1958, as the eradication program was underway, only Louisiana and Mississippi listed the red imported fire ant among the top twenty insect pests.[182]

Conclusion

This chapter traced the construction of the red imported fire ant's image, arguing that it reflected not just the ant's biology nor just the ideas of those who interacted with it, but a combination of the two. Natural history and human history intertwined, although in this case they did not have equal power. Driven by changes within the USDA, the image of the ant diverged significantly from the ant's behavior.

When the ant first reached North America, it confronted a culture preadapted to view it with suspicion. Changes in agriculture, the ecology of insects, and the orientation of myrmecologists combined to make ants into a symbol of social decay. *Solenopsis invicta* then fell under the gaze of a number of different people with different ideas, different ways of working, and different intellectual commitments. Control entomologists studied the insect to assess the damage that it did and find ways to curb its spread; E. O. Wilson studied the ant as an economic entomologist and also as a way of exploring evolutionary theory; journalist Bill Ziebach blamed the ant for ruining quail hunting, an important part of southern culture, and for ruining southern agriculture. All of these people studied the same insect and no one fabricated data, but because they looked at the ant from varying perspectives, they saw different aspects of the insect. While they all agreed that the ant damaged some crops and some pastures, they could not agree upon the severity of the problem posed by the ant nor the proper remedy. Some saw the ant as a nuisance; others saw it as a significant pest; and Wilson considered it among the worst ant pests in the Western hemisphere. Some recommended eradication, some small-scale control, and some proposed doing nothing. In 1957, the Arkansas entomologist Charles Lincoln said, "Let us either try to eradicate the beast, or else learn to live with it."[183] The ant could be eradicated—since chemical insecticides seemed so benign and eradication so easily accomplished—but it could also be left alone.

The same year that Lincoln offered his suggestion, the Plant Pest Control Division of the USDA and its allies cut through the ambiguity and insisted that the ant was a major pest that needed to be eradicated. This decision rested not on new information about the insect, but reflected changes in the department's organization. A new, unproven

---

182. *Cooperative Economic Insect Report* 8 (May 1958): 440–70.
183. C. Lincoln to J. W. White, 5 April 1957, fire ant files (1957), PPC papers.

division, with new and audacious goals, had come into bureaucratic being, and its officials chose to eradicate the ant in order to prove the validity of those goals and the power of their bureaucracy. Convinced of the potency and safety of chemical insecticides, they sorted through the earlier reports, accepted the ones that fit their agenda, and ignored the rest. The ant played a role in this bureaucratic evolution—it provoked requests for aid from southerners, giving the USDA a mandate, and the ant's biology informed Wilson's study, on which the USDA relied. But, the ant alone does not explain the advent of the eradication program—nor does the PPC's ambitions. Natural history and human history combined, although not in equal measure: the decision to eradicate the ant rested more on the agency's internal politics than any particular characteristic of the ant.

Chapter Three: Fire Ants, from Savage to Invincible, 1957–1972

What wonder strikes the curious, while he views
The black ant's city, by a rotten tree
Or woodland bank! In ignorance we muse:
Surely they speak a language whisperingly,
Too fine for us to hear; and sure their ways
Prove they have kings and laws, and that they be
Deformed remnants of the fairy-days.

A. S. Byatt, *Angels and Insects*

In 1969, William Brown complained, "During the past half-century there have been surprisingly many popular books published on ants. Most of these seem to have been based largely on a single compendium, William Morton Wheeler's *Ants*. In 1910, Wheeler's book was a fresh and authoritative source, but it has been reprinted four times since then without revision. An enormous amount has been learned about ants since 1910, but the popular literature has failed to reflect accurately this gain in knowledge."[1] Inspired, perhaps, by the disdain that Brown felt toward his scientific predecessors, the diagnosis was still apt: Wheeler's ideas defined the popular understanding of ants.[2] But, prevalent as Wheeler's ideas were, there were other assessments of ants in which they were not so vile as often portrayed. Brown's generation, busily deconstructing the received wisdom that ants were innately communists, contributed to these contrary beliefs. "Each individual ant has an ability to learn and remember and respond to situations. It has, moreover, an ability to do something off its own bat," wrote the biologist Derek Morley in 1953.[3] (In 1967, Creighton rewrote his article on ants for the *Encyclopedia Britannica*, dropping the section on ants

1. W. L. Brown, "The Lore of the Ant," *Audubon* 71 (January 1969): 86.
2. For Brown's relationship with his myrmecological ancestors, see J. B. Buhs, "Building on Bedrock: William Steel Creighton and the Reformation of Ant Systematics," *Journal of the History of Biology* 33 (2000): 55–64.
3. D. W. Morley, *The Ant World* (London: Penguin Books, 1953), 171. For more examples, see E. O. Wilson, *The Insect Societies* (Cambridge: Harvard University Press, 1971), 317–19. For a consideration of similar changes throughout biology, see G. Mitman, "Defining the Organism in the Welfare State: The Politics of Individuality in American Culture, 1890–1950," in *Biology as Society, Society as Biology*, ed. S. Maasen, E. Mendelsohn, and P. Weingart (Dordrecht: Kluwer Academic Publishers, 1995), 249–78.

as pests to make room for some of this new research.)[4] At about the same time that Brown bemoaned Wheeler's continuing influence, Marvel comics introduced the super hero Ant Man (1962) while Hanna-Barbera Studios produced the cartoon Atom Ant (1965)—in which a mutated insect did not threaten the nation, but defended it.[5] Ants and atoms could be forces of civilization. There were as well those who continued to view ants through the lens of Christianity. In the 1940s the pastor Joshua Stauffer, for example, wrote a series of essays for young people using ants to explore Christian ideals.[6]

Hence, a continuum of views about insects, ants, and, specifically, the imported fire ant proliferated in the wake of *Solenopsis invicta*'s irruption across the South. One line of thought culminated in the interpretation advanced by the U.S. Department of Agriculture (USDA)—that the red imported fire ant was a menace to the nation's agriculture that needed to be eradicated. This view was probably the dominant one, but others existed along that continuum, some characterizing the ant as merely a nuisance (as some early USDA reports did) and others emphasizing the ant's beneficial attributes.

Not only was the ant undergoing reevaluation in the years after the eradication campaign began, the status of the chemicals used to kill the insect also fluctuated. Many still held insecticides to be safe and useful tools. Others, however, began to question their safety, and some stated outright that the chemicals posed a serious danger to the environment, wildlife, and humans. In 1962, Rachel Carson published *Silent Spring*, in which insecticides figured as a symbol of human hubris. Although insecticides were chemicals, composed of atoms, just like the imported fire ant, and subject to natural laws—affecting the physiology of plants and animals in predictable and explicable ways—Carson argued that they were unnatural. As skepticism about the safety of insecticides increased, the ant began to look less menacing. And although it evolved thousands of miles from North America, many came to see the insect as part of the South's ecosystem. Manufactured chemicals were not part of nature's order, but an insect brought to Mobile, Alabama by ships and spread through trade was.

4. W. S. Creighton to Editors, *Encyclopedia Britannica*, 29 May 1967, William Steel Creighton papers, an unprocessed box of material incorporated in the uncatalogued E. O. Wilson papers, Library of Congress, Washington, D.C. (hereafter Creighton papers).

5. On Ant Man, see L. Daniels, *Marvel: Five Decades of the World's Greatest Comics* (New York: Abrams, 1991), 97–98. On Atom Ant, see "Don Markstein's Toonopedia," available at www.toonopedia.com/atomant.htm, 19 October 2002 (accessed 24 June 2003).

6. J. Stauffer, *The Ants Are a People* (Westfield, Ind.: Gospel Minister, 1947).

This chapter considers the alternative interpretations of the ant and the insecticides in wake of the eradication campaign. Who opposed the USDA's eradication campaign? How did they account for the ants, the insecticides, and the eradication ideal? And how did their ideas about the ant reflect their position in American society and the natural world that they were trying to understand?

"Doing More Good Than Harm"

The view that the red imported fire ant might be a positive addition to the North American ecosystem—or was, at least, no more than a minor nuisance that could be controlled where it caused problems, but did not need to be eradicated—had its roots in the work of a coterie of southern university entomologists. In the years after the close of the Spring Hill laboratory, federal entomologists quit studying the insect, but the South's university entomologists did not. The Alabama Polytechnic Institute (renamed Auburn University in 1960) and Mississippi State University (MSU) sponsored small-scale investigations throughout the 1950s.[7] As interest in the ant peaked in 1956 and early 1957, the schools invested more resources while entomologists at Louisiana State University (LSU) also began to study the insect.[8]

For the most part, these entomologists sought a clearer biological understanding of the ant, and shared a disdain for the eradication ideal set forth by the USDA's Plant Pest Control Division (PPC) and its focus on chemical insecticides to the exclusion of all other insect control methods.[9] LSU's Leo Newsom already doubted the possibility of eradicating insects when he heard Clay Lyle speak in 1946, and acquaintance with the fire ant program did nothing to alter his opinion.[10]

7. For MSU's work, see H. B. Green, "Biology and Control of the Imported Fire Ant in Mississippi," *Journal of Economic Entomology* 45 (1952): 593–97. For Auburn's work, see "The Alabama Polytechnic Institute and the Imported Fire Ant," 21 March 1957, box 2, Ralph B. Draughon papers, RG 107, Auburn University Archives (hereafter Draughon papers).

8. For a description of LSU's research on the imported fire ant, see G. Stromeyer, "Unraveling the Secrets of the Fire Ant," *American Mercury* 91 (July 1960): 121–24.

9. For examples of this disdain, see B. A. Porter to J. L. Horsfall and H. M. Harris, 13 September 1949, file 18/27, Entomological Society of America papers, Iowa State University, Park Library; J. A. G. Rehn to J. C. Bradley, 26 March 1951, Entomological Society of America file, James C. Bradley papers, Cornell University archives; M. H. Hatch, "Entomology in Search of a Soul," *Annals of the Entomological Society of America* 47 (1954): 377–87; H. H. Ross, "Free Trade," *Bulletin of the Entomological Society of America* 1 (1955): 6–7, 15.

10. On Newsom's reaction to Lyle's speech, see L. D. Newsom, "H. H. Schwardt: Scientist, Teacher, Friend," in *Addresses Given at the Dedication of the H. H. Schwardt Laboratory for Research on Insects Affecting Man and Animals* (Ithaca, N.Y.: n.p., 1976), 5. For his continued skepticism, see L. D. Newsom, "Eradication of Plant Pests—Con," *Bulletin of the Entomological Society of America* 24 (1978): 35–40.

MSU's Ross Hutchins and H. B. Green stated, "The very nature of the ant and its peculiar habits make it a difficult pest to eradicate."[11] And Frank S. Arant, the head of Auburn's entomology department, agreed. In 1958, he told the USDA, "I am anxious not to be on record at this time supporting an all-out eradication program."[12] The decision by the university entomologists to study the imported fire ant was in part a response to public calls for aid in dealing with the invading insect: they wanted to help southern citizens. But, I contend, it was also a coun- teroffensive against the PPC and its eradication program. Newsom, Hutchins, and others concerned about the proliferation of the eradica- tion ideal and reliance on chemical insecticides no doubt agreed with William Brown when he said, "We need biologists and not administra- tors and not squirt-gun people in charge of our control programs."[13]

The contours of the ant's new image came into focus in early 1957, when Alabama sent Auburn's Kirby Hays to South America to observe the ant in its land of origin and learn how locals dealt with the pest.[14] A recent recipient of a Ph.D. from the University of Michigan, Hays had studied mosquitoes for the army during the Korean conflict and had also done some work in Puerto Rico.[15] Now, under the aegis of the State of Alabama, his goal was to identify possible predators and para- sites that might be imported into North America to control the fire ant. Hays was only in South America for two weeks, not enough time to conduct a rigorous study of the ant's natural history, so he supple- mented his own observations by interviewing entomologists and farm- ers. He found that the ant had several predators, but concluded it was unlikely that they could be effective in controlling the insect in the American South. More intriguing, however, was Hays's discovery that the people with whom he spoke did not consider the ant a pest at all, even in places where the density of its nests approached one hundred mounds per acre. The ant ate other insects that farmers considered pests; they welcomed the ant as an ally.[16] "All specialists consider these

11. R. E. Hutchins and H. B. Green, "Research Staff Begins Imported Fire Ant Study," *Missis- sippi Farm Research* 20 (May 1957): 1.

12. C. C. Fancher to E. D. Burgess, 12 May 1958, fire ant file, box 197, Plant Pest Control Division papers, RG 463, National Archives and Records Administration, College Park, Md. (here- after PPC papers). See also F. S. Arant to E. V. Smith and J. L. Lawson, 12 March 1957; and W. A. Ruffin to R. B. Draughon, 29 April 1958, both in box 2, Draughon papers.

13. Quotation from F. Graham, *Since Silent Spring* (Boston: Houghton Mifflin, 1971), 23.

14. J. E. Folsom to R. B. Draughon, 1 April 1957, box 2, Draughon papers.

15. For biographical data on Hays, see "Fire Ant Team Has Expert on Subject of Insects Along," 1 May 1957, fire ant files (1957), PPC papers.

16. For Hays's observations, see K. L. Hays, "The Present Status of the Imported Fire Ant in Argentina," *Journal of Economic Entomology* 51 (1958): 111–12.

ants to be beneficial . . . and cannot understand the concern of North American farmers," Hays reported to the James Folsom, Alabama's governor.[17]

Back home, Hays's brother Sydney, a candidate for a master's degree at Auburn, found that the ant behaved in North America much as Kirby found it did south of the equator. Sydney Hays marched through fields near Auburn, looking for fire ant mounds, tearing them open, and pulling out the food that the insects had gathered. The ant, he learned, was omnivorous, but mostly it ate other insects. "We saw the ants returning [to their nests] with aphids, larvae of different insects, small spiders, and even hard bark beetles," he said. "The ants brought the insects back to the tunnel openings, cut them apart, ate the soft parts, and left the hard pieces on the ground."[18] He also watched the ant harvest maggots from cow manure. In the laboratory, Hays presented colonies with a variety of different foods, from seeds to insects. Many resorted to cannibalism before vegetarianism.[19] The ant preferred to eat other insects—including those that were undoubtedly pests. In one test, Hays caged two hundred boll weevils near a fire ant mound. The ants ate all two hundred in a single evening.[20]

These studies were not without their detractors. The USDA dismissed them, in large part because the investigations of E. O. Wilson seemed to invalidate them: the red imported fire ant is a mutant, the Harvard myrmecologist had said, from which the USDA had concluded that the ant's behavior in South America had no bearing on its behavior north of the equator. The insect's unusual genetic programming made it a pest here.[21] Marion Smith and William Creighton had their own reasons for doubting the work coming out of Auburn. They learned that Hays and other Auburn biologists had consistently confused the southern fire ant, *Solenopsis xyloni*, for the import.[22] No wonder the Auburn

17. "The Present Status of the Imported Fire Ant, *Solenopsis saevissima richteri* Forel, in Argentina," box 2, Draughon papers.

18. G. Stevenson, press release, 21 February 1958, box 2, Draughon papers.

19. S. B. Hays, "The Food Habits of the Imported Fire Ant, *Solenopsis saevissima richteri* Forel, and Poison Baits for its Control" (Master's thesis, Alabama Polytechnic Institute, 1958); S. B. Hays, "Food Habits of *Solenopsis saevissima richteri* Forel," *Journal of Economic Entomology* 52 (1959): 455–57. See also K. L. Hays, "Ecological Observations on the Imported Fire Ant, *Solenopsis saevissima richteri* Forel, in Alabama," *Journal of the Alabama Academy of Science* 30 (1959): 14–18.

20. G. Stevenson, press release, 21 February 1958, box 2, Draughon papers.

21. PPC to J. L. Abbot, 27 June 1957, fire ant files (1957), PPC papers. See also L. J. Padget to E. D. Burgess, 10 June 1960, fire ant files (1960), PPC papers.

22. On the taxonomic confusion, see M. R. Smith to P. Oman, 10 November 1957, fire ant files (1957), PPC papers (quote); W. S. Creighton to W. H. Grimes, 19 November 1957; W. S. Creighton to W. H. Grimes, 23 November 1957; M. R. Smith to W. S. Creighton, 9 December 1957; M. R. Smith to W. S. Creighton, 19 February 1969, all in Creighton papers.

researchers found that the insect caused few problems! They had been brought to grief by tampering with what Creighton had called the "taxonomic duds" that made studying *Solenopsis* so difficult, and accidentally studied a native ant instead of the invader.

This taxonomic confusion, though, was not widely known, and soon new evidence emerged that seemed to support the Hays brothers' conclusions and undermined the USDA's credibility, especially among those predisposed to distrust the department. At LSU, for instance, Newsom and his colleagues learned that the ant was an important predator of the sugar cane borer. When the USDA sprayed heptachlor onto sugar cane fields, the ant's population plummeted, while the borer's population exploded. The biologists reasoned that the ant had replaced other insects as the borer's main predator and now *Solenopsis invicta* kept the pest population under control.[23] The results were so striking that some sugar cane farmers considered suing the USDA for killing the ant and ruining their crop.[24] Meanwhile, H. B. Green concluded that the ant increased the fertility of the southern soil. Casting the remnants of devoured insects into large piles recycled nutrients while the ant's tunnels aerated the ground.[25] Grass grew richly around fire ant nests, he noted, and not because the insect scared off cattle: Arant, Kirby Hays, and their colleague Dan Speake pointed to their findings that "[c]attle and sheep graze over the mounds and even lie down near them. Newborn livestock is rarely if ever killed."[26]

As a result of these discoveries, the southern university entomologists developed an interpretation at odds with the one espoused by the USDA. They did not all agree on the severity of the problems caused by the ant—or if the ants caused significant problems at all—but opinions clustered around the belief that the ant was at most a nuisance. As one USDA entomologist noted, "There seems to be a tendency, on the part of some research men, to minimize the damage that fire ants do to both animals

23. W. H. Long, E. A. Cancienne, E. J. Cancienne, R. N. Dopson, and L. D. Newsom, "Fire Ant Eradication Program Increases Damage by the Sugarcane Borer," *Insect Conditions in Louisiana* 1 (1958): 10–11; W. H. Long, E. A. Cancienne, E. J. Cancienne, R. N. Dopson, and L. D. Newsom, "Fire Ant Eradication Program Increases Damage by the Sugarcane Borer," *Sugar Bulletin* 37 (1958): 62–63.

24. J. C. Caldwell to H. S. Peters, 12 August 1959, National Audubon Society papers, New York Public Library (hereafter NAS papers).

25. *The Imported Fire Ant in Mississippi*, Bulletin 737 (State College: Mississippi State University, 1967). For more on Green's work, see H. B. Green, "Biology and Control of the Imported Fire Ant in Mississippi," *Journal of Economic Entomology* 45 (1952): 593–97.

26. F. S. Arant, K. L. Hays, and D. W. Speake, "Facts about the Imported Fire Ant," *Highlights of Agricultural Research* 5 (Winter 1958): 12.

and plants."[27] Sydney Hays, for example, told the press, "The fire ant is not guilty of many charges," but he did not want his research to be used to discourage the control of the ant. The insect remained a major nuisance, stinging painfully and building mounds that interfered with farming operations.[28] Others conceded less. Arant claimed, "Damage to crops is rare. It is possible that an occasional animal has been killed, but we know of no authentic record of such happenings." And Hutchins said, "The imported fire ant is not a damaging economic insect."[29] Indeed, Hutchins insisted, the ant was so well integrated into the southern ecosystem that it was essentially a native insect.[30] But for these researchers, as for Sydney Hays, these findings did not mean that the ant was harmless: even native insects were irksome at times, and the imported fire ant's large, hard mounds and tendency to sting qualified the bug as a nuisance. Yet being a nuisance is not the same as being an agricultural pest, or a threat to public health. Arant and Hutchins thought that the insect should be controlled ("Control of the imported fire ant is essential on certain croplands, pastures, haylands, and lawns," Arant said), but dismissed both the possibility and the need for eradication—a very different response to the ant (and insect problems generally) than that offered by Clay Lyle, Irma Fortune, and the PPC (fig. 10).[31]

Leo Newsom pushed the data the hardest, transforming the ant from nuisance to ally. His research and the research of others, he claimed, "[p]roved beyond any doubt that the imported fire ant is not an important pest of agriculture in the United States."[32] He admitted that the ant's mound might be problematic at times and that the sting was painful—although he allowed himself to be stung over 270 times with no more ill-effects than some pustules that eventually waned—but, despite these hassles, he told a reporter, "The fire ant may be doing more good than harm."[33] He called attempts to eradicate it "idiotic."[34]

27. C. S. Lofgren to A. W. Lindquist, 7 February 1958, insects affecting man and animals branch fire ant file, box 84, Entomology Research Division: General Correspondence, 1954–1958, 5 (UD), Agricultural Research Service papers, RG 310, National Archives and Records Administration, College Park, Md. (hereafter ARS papers).

28. G. Stevenson, press release, 21 February 1958, box 2, Draughon papers.

29. Arant's quotation is from F. Bellinger, R. E. Dyer, R. King, and R. B. Platt, "A Review of the Problem of the Imported Fire Ant," 1965, Creighton papers. Hutchins's quotation is from R. E. Hutchins to P. Charam, 14 July 1964, imported fire ant file, box 247, PPC papers.

30. R. E. Hutchins to P. Charam, 17 July 1964, imported fire ant file, box 247, PPC papers.

31. F. S. Arant, "Status of the Imported Fire Ant," 30 June 1958, fire ant file, box 197, PPC papers.

32. L. D. Newsom to P. Charam, 17 July 1964, imported fire ant, box 247, PPC papers.

33. Quotation and the number of times Newsom was stung from "Mirex Muddle: Entomologist Says Fire Ant Control Plan 'Idiotic,' " n.d., Murray S. Blum fire ant materials, in author's possession (hereafter Blum papers).

34. Ibid.

Figure 10. "The Imported Fire Ant in Mississippi." Some Mississippi entomologists opposed the USDA's eradication program although they were no friend of the ant. They wanted to kill the insect where it caused problems, but doubted the efficacy of the federal program. This picture, from the cover of a Mississippi state pamphlet on the ant (written by H. B. Green), shows a woman next to a large fire ant nest, the kind that interfered with tractors and made Mississippi entomologists intent to control the ant. (From "The Imported Fire Ant in Mississippi," Bulletin 737, February 1967, Mississippi State University.)

The ant that Newsom, Arant, Hutchins, and Green saw looked very different than the one described by E. O. Wilson, Bill Ziebach, and the PPC (see chapter 2). Part of the reason for this difference can be found in the varying methods used to study the insect. Wilson, sympathetic to the kind of entomology practiced by the Auburn scientists (as opposed to the one advocated by the PPC), although he questioned their results, noted that studying the insect in the laboratory might have led Sydney Hays astray: the ant might attack crops and wildlife only under certain conditions, but because those conditions were unknown, they could not be replicated in controlled situations.[35] Wilson's critique seems a reasonable description of the limitations of the new view of the ant. The South's entomologists saw only a slice of the ant's behavior—in the laboratory or in areas where its population had crested and began to decline. They did not rely heavily on farmer surveys, insulating themselves from the exaggerated tales that bedeviled USDA scientists a decade before, but also shielding themselves from seeing the damage

35. E. O. Wilson to P. Charam, 14 July 1964, imported fire ant file, box 247, PPC papers.

that the ant did, evidence that persuaded Wilson that the insect could be a significant agricultural pest. The ant's seemingly docile behavior was an artifact of the way the southern university entomologists studied it. Newsom saw the insect devour sugar cane borers, but not eat okra or potatoes or quail, and so he assumed that the ant caused fewer problems than it in fact did.

The differences in perspective between southern entomologists and their forerunners, though, are not attributable only to different methods. In the early 1960s, when he became convinced that the reports from the South's entomologists were substantially correct (although he continued to dismiss the work coming out of Auburn), E. O. Wilson suggested that the ant had changed its behavior in ways that accounted for the different perspectives.[36] In the early years of the invasion, the insect's population density had been high, and to survive the opportunistic ant had eaten crops, possibly wildlife, too. Wilson had seen the damage with his own eyes and could not discount it. But since then, the ant's population had declined in some areas, and, no longer pressed for calories, the insect turned its attention to its preferred food—other insects. Of course, since the ant is opportunistic, and because there are always new areas to exploit—as new suburbs are built, or new cattle pastures made—the insect could still cause problems, sometimes severe ones, even in places it had long occupied. But generally, it would be more tame—seem more a native insect, as Hutchins said—in places that it had invaded years before. Not everyone agreed with Wilson's assessment: Creighton said that he made an "ass of himself" just suggesting it, arguing that Wilson had invented the theory to save his own observations from skepticism, when other researchers could not confirm them. There is evidence, however, to suggest that Wilson might have been right.[37]

In the late 1950s, Auburn entomologists noted that, due to competition among colonies, the number of imported fire ant nests in some places dropped from over one hundred per acre to less than forty.[38] Two score colonies on a single acre was still enough to cause problems, but it also may have meant that the ant could find enough insects to eat and therefore could ignore crops and vertebrates—just

36. E. O. Wilson to P. Charam, 14 July 1964, imported fire ant file, box 247, PPC papers. See also F. Bellinger, R. E. Dyer, R. King, and R. B. Platt, "A Review of the Problem of the Imported Fire Ant," 1965, Creighton papers.

37. Creighton's quotation is from W. S. Creighton to R. E. Gregg, 9 March 1968, Creighton papers.

38. F. S. Arant, K. L. Hays, and D. W. Speake, "Facts about the Imported Fire Ant," *Highlights of Agricultural Research* 5 (Winter 1958): 13.

as in the 1940s Irma Fortune had surmised that the ants only attacked corn when meaty, greasy, food was unavailable.[39] Arant thought that Wilson's argument was reasonable, saying, "Apparently these (early) reports (of widespread damage to crops, livestock and newborn animals) were erroneous or the ant has changed its habits."[40] The suggestion that the ant changed its feeding habits is only speculation, but it explains how the insect elicited different reactions at different times and in different places, and the pattern of those responses.

By the early to mid-1960s the belief that the ant was not a pest seems to have become conventional wisdom for many entomologists. In 1965, for example, scientists reviewing the fire ant problem in the *Bulletin of the Georgia Academy of Science* wrote, "The imported fire ant should not now be considered as a significant *economic* pest."[41] Two years later, the National Academy of Sciences convened a panel to review the disparate reports on the ant. The panel concluded that the claim that the insect "significantly affects land values, the labor market, agricultural productivity, or national health and welfare was not effectively demonstrated."[42] Even Wilson had changed his mind. In 1964, he told the USDA that he was convinced that the insect was no longer a significant agricultural pest.[43] This idea had traction, I think, because it helped to substantiate the contention that biologically inclined entomologists understood the insect and their science better than those who favored chemical solutions and advanced the eradication ideal. Wilson, for example, testified before a congressional subcommittee in 1963, "The research on this species, other than routine scouting and insecticide tests, was quite trivial. . . . Had but a small fraction of the 2.4 million dollars been devoted to basic research, there would now be solid achievements to build upon."[44] He and Brown, with another Harvard entomologist, blasted the USDA's eradication

---

39. I. Fortune, "The Biology and Control of *Solenopsis saevissima* variety *richteri* Forel," (Master's thesis, Mississippi State College, 1948), 7.

40. F. Bellinger, R. E. Dyer, R. King, and R. B. Platt, "A Review of the Problem of the Imported Fire Ant," 1965, Creighton papers.

41. F. Bellinger, R. E. Dyer, R. King, and R. B. Platt, "A Review of the Problem of the Imported Fire Ant," 1965, Creighton papers (emphasis in original).

42. "Appraisal of Programs for Fire Ant Control," 1 March 1967, Committee on Fire Ants, central file, National Academy of Sciences (Biology and Agriculture) papers, National Academy of Sciences archives, Washington, D.C.

43. Compare E. O. Wilson to R. L. Burlap, 11 January 1959, fire ant file, box 213, and E. O. Wilson to P. Charam, 14 July 1964, imported fire ant file, box 247, both in PPC papers.

44. "Statement Read before the U.S. Senate Subcommittee on Reorganization and International Organizations," 8 October 1963, Creighton papers.

program.[45] They wanted more research, a point that Arant continually pressed.[46]

Newsom best illustrates my point that claims of the ant's benefi-cence were part of an argument about how best to practice entomology. Although he said that the ant might be doing more good than ill, he also admitted if he "had the power to snap my fingers and get rid of the imported fire ant today without causing any problem to the environ-ment, I would do it."[47] But why eradicate an ant that was more benefi-cial than not? I think that Newsom exaggerated claims about the ant's beneficence because he was frustrated with the eradication program. He experienced more pressure from eradication advocates than any of the other university entomologists. Creighton remembered one meet-ing in which Newsom and a proponent of the eradication program got into a shouting match that degraded to what Creighton called "the son of a bitch level."[48] And when Newsom tried to protest the USDA's actions, he continually found himself thwarted by those associated with Louisiana's agricultural industry, which supported the eradication effort.[49] It was in the face of this resistance that Newsom most insis-tently pointed to the ant's good qualities, as a way to prove how wrong-headed his opponents were.

The belief that the ant was only a nuisance, and maybe even ben-eficial, also gained a foothold in the popular culture in the latter part of the 1950s, though it remained overshadowed by the USDA's pro-nouncements. The work of Kirby and Sydney Hays, for example, received a good deal of coverage in the early months of 1957, and influ-enced a number of the USDA's critics.[50] On a national level, *Reader's*

45. W. L. Brown, E. O. Wilson and H. W. Levi to E. T. Benson, 8 December 1957, complaints-2-fire ant spray program file, box 195, PPC papers. See also "3 Experts Protest Fire Ant Program," *New York Times*, 27 December 1957, 11.

46. For Arant's insistence on more research, see "Imported Fire Ant Research Conference," 23–24 September 1958, Entomology Research Division: General Correspondence, 1954–1958, 5 (UD); F. S. Arant to E. F. Knipling, 28 April 1959, box 1, man and animals imported fire ant file, Entomology Research Division: Entomology Director's Correspondence, 1959–1965, 1055 (A1), both in ARS papers.

47. "Statement of Leo Dale Newsom," 10 April 1974, Newsom, 4/30/74 file, box 63, Environ-mental Defense Fund papers, MS 232, State University of New York, Stony Brook, N.Y. (hereafter EDF papers). See also L. D. Newsom to L. J. Padget, 24 March 1959, in "Comments on the GAO Conference Memorandum," fire ant file, box 128, PPC papers.

48. W. S. Creighton to M. R. Smith, 24 February 1969, Creighton papers.

49. On Newsom's difficulties, see R. N. Dopson to C. C. Fancher, 12 June 1961, fire ant-1 file, box 93; D. L. Pearce to P. Charam, 20 July 1964, imported fire ant file, box 247, both in PPC papers. See also R. Van den Bosch, *The Pesticide Conspiracy* (New York: Doubleday, 1978), 61, 63.

50. The publicity is described in W. O. Owen to C. C. Fancher, 13 May 1957, fire ant files (1957), PPC papers.

Digest reconsidered its opinion of the ant. In September 1957, as Congress contemplated funding the fire ant eradication program, the magazine warned, "A formidable army of South American fire ants has invaded the United States," and named the insect "one of the most conspicuous nuisances ever to threaten U.S. farmers and the citizenry at large."[51] Two years later, it conceded that the "fire ant is not a serious crop pest; it may not be a crop pest at all."[52] *Outdoor America* and *Florida Wildlife* offered similar revisions of previous assessments of the ant.[53] In 1982, the *Miami Herald* reported, "The little red beastie that Southerners love to hate is courted, protected and even worshipped in the swampy Louisiana lowlands above and below New Orleans, where America's largest sugar cane crop is grown."[54] The ant did not inspire a statue, but, as the boll weevil before it, the insect found a corner of the South where it was celebrated.

Wildlife, Insecticides, and the Eradication Ideal

A diverse group often referred to as "conservationists" represented another bloc of opposition to the USDA's eradication program. Composed of hunters; fishermen; outdoor enthusiasts; wildlife biologists; advocacy groups such as the Conservation Foundation, the National Audubon Society, the National Wildlife Federation and their state affiliates; and, at the federal level, the U.S. Fish and Wildlife Service, the numbers of conservationists were substantial: about 25 million Americans bought hunting or fishing licenses in 1954.[55] Variously influenced by the critique of the eradication program on the part of university entomologists, the conservationists, for the most part, insisted that the insecticides used to kill the imported fire ant posed a greater threat to the South's fauna than the invading insect.

Over the years since the 1930s, when Herbert Stoddard first had observed a native species of fire ant eating quail in southern Georgia and northern Florida, wildlife biology had become a full-fledged scientific discipline, with professional journals as a vehicle for research and departments in the nation's land-grant universities, such as Auburn,

---

51. A. Rankin, "The Great Fire-Ant Invasion," *Reader's Digest* 71 (September 1957): 74–77.

52. R. S. Strother, "Backfire in the War against Insects," *Reader's Digest* 74 (June 1959): 64–69.

53. J. W. Penfold, "The Fire Ant Revisited," *Outdoor America* 24 (1959): 5–7; compare J. Wheeler, "The Fire Ant," *Florida Wildlife* 11 (March 1958): 28–29, 40—41, with S. Ehlers, "The Incredible Imported Fire Ant," *Florida Wildlife* 27 (3 August 1973): 24–27.

54. B. Rose, "Friend or Foe? Ants That Farmers Love to Hate Find Friends in Louisiana," *Miami Herald*, 14 June 1982, 1A.

55. On the number of hunters and fishermen, see *National Survey of Fishing and Hunting*, U.S. Fish and Wildlife Service circular no. 44 (Washington, D.C.: U.S. Fish and Wildlife Service, 1955).

MSU, and LSU.[56] Stoddard's work was one of the foundations of the new discipline.[57] In the past, wildlife biologists had spent most of their time controlling predators; Stoddard prompted them to reconsider this focus.[58] His research found that predators had little or no effect on game populations. If he controlled the fire ant, then more quail would succumb to cold or disease or some other predator.[59] The key to maintaining an abundant game population, then, was not killing predators, but creating a habitat that supported as much wildlife as possible. Aldo Leopold, the architect of modern wildlife biology, built on this idea, explaining that game management was "the art of making land produce sustained annual crops of wild game for recreational use"; it was "best understood by comparing it with other land-cropping arts"—like agriculture.[60] Indeed, wildlife biologists shared with officials in the USDA a belief in the conservation ethic, what historian Samuel Hays calls the "gospel of efficiency." Nature was a set of resources. It was the task of scientists to find the most efficient way of utilizing the resources.[61]

Despite this shared philosophy, farmers of agricultural crops and farmers of game animals sometimes clashed, for although both agriculture scientists and wildlife biologists tended crops, their crops had different, sometimes contradictory, needs. The USDA, for example, did not like wastelands and wanted them converted to fields and pastures, while officials with the U.S. Fish and Wildlife Service (USFWS) blanched at the thought. Quail lived on those so-called wastelands. They were not wastes at all; they were productive.[62] The use of insecticides (and the

56. T. R. Dunlap, *Saving America's Wildlife: Ecology and the American Mind, 1850–1990* (Princeton: Princeton University Press, 1988), 70–79.

57. On the importance of Stoddard's work, see C. Meine, *Aldo Leopold: His Life and Work* (Madison: University of Wisconsin Press, 1987), 264.

58. On wildlife biology's concern with controlling predators, see T. R. Dunlap, *Saving America's Wildlife: Ecology and the American Mind, 1850–1990* (Princeton: Princeton University Press, 1988), 34–64.

59. For the development of Stoddard's thought, see H. L. Stoddard, *The Cooperative Quail Study Association: May 11, 1931–May 1, 1943* (Tallahassee: Tall Timbers Research Station, 1961); E. V. Komarek, "Comments on the 'Fire Ant Problem,'" *Proceedings of the Tall Timbers Conference on Ecological Animal Control by Habitat Management* 7 (1978): 1–9.

60. Quoted in T. R. Dunlap, *Saving America's Wildlife: Ecology and the American Mind, 1850–1990* (Princeton: Princeton University Press, 1988), 76.

61. S. P. Hays, *Conservation and the Gospel of Efficiency: The Progressive Conservation Movement, 1890–1920* (Cambridge: Harvard University Press, 1959).

62. For this debate, see B. T. Shaw to O. L. Meehan, 25 April 1956; J. L. Farley to B. T. Shaw, 17 May 1956, both in pesticides 1949–1960 file, box 24, Wildlife Research Correspondence, 254 (UD), U.S. Fish and Wildlife Service papers, RG 22, National Archives and Records Administration, College Park, Md. (hereafter USFWS papers).

imported fire ant) also divided the conservation philosophy of agricultural officials from the one advocated by wildlife biologists. Proponents of the new insecticides and the eradication ideal championed their work as "conservation—no matter how you looked at it," in W. L. Popham's words. He told the Audubon Society, "In many respects . . . our long-range objectives are quite compatible."[63] Some wildlife might die, Popham acknowledged, since insecticides were occasionally toxic to vertebrates, but the benefits of insecticide use outweighed the risks. Theoretically, eradication meant that an area only had to be treated once or twice, not annually, meaning that less insecticide could be used than in years past, saving money and exposing game animals to less poison as well. Since bugs would be driven extinct, another entomologist explained, crops could be grown densely, without fear of insect attack, leaving more room for wildlife.[64] And the eradication of the imported fire ant would help game populations by vanquishing a terrible predator.

Wildlife biologists were not so sanguine about the insecticides and had varying opinions about the importance of the ant. In the years since World War II, insecticides had come under increasing scrutiny as a threat to animals. When, in the late 1940s, Bill Ziebach led Operation Ant and the Alabama Department of Conservation distributed free chlordane, not much had been known about the effect of chemical pesticides. The department, for instance, thought that DDT had virtually no effect on quail.[65] Increasingly, however, research showed that opinion to have been naïve. The USFWS, working with the USDA, demonstrated that one pound of DDT per acre killed significant numbers of fish, and five pounds per acre wreaked havoc on forest animals.[66] Experiences with the Dutch elm disease program in the Midwest and

63. "Conservation through Pest Control," 3 June 1958, information—speeches file, box 82, Entomology Research Division: General Correspondence, 1954–1958, 5 (UD), ARS papers (first quotation); W. L. Popham to S. H. Ordway Jr., 26 December 1957, regulatory crops-1-fire ants file, box 752, General Correspondence, 1954–1966, 1 (UD), ARS papers (second quotation). On the USDA's opinion more generally, see D. G. Hall, "Food, Wildlife, and Agricultural Chemicals," *Conservation News* 17 (1 July 1952): 1–4.

64. G. C. Decker, "Pesticides Relationship to Conservation Programs," *National Agricultural Chemicals Association News and Pesticide Review* 17 (May 1959): 7.

65. R. Allen to M. E. Hicks, 27 June 1949, conservation file, SG 9977, Alabama Department of Conservation papers, Alabama Department of Archives and History, Montgomery, Alabama (hereafter ALDoC papers).

66. C. Cottam and E. Higgins, "DDT and its Effects on Fish and Wildlife," *Journal of Economic Entomology* 39 (1946): 44–52; J. L. George and R. T. Mitchell, "The Effects of Feeding DDT-treated Insects to Nesting Birds," *Journal of Economic Entomology* 40 (1947): 782–89; J. L. George and W. H. Stickel, "Wildlife Effects of DDT Dust Used for Tick Control on a Texas Prairie," *American Midland Naturalist* 42 (1949): 228–37.

the Gypsy moth program in the Northeast confirmed these fears: wildlife exposed to the insecticides did die, although no one was certain how badly the USDA's eradication programs hurt wildlife. Insecticides may have been a boon for agriculture, but they seemed a bane for game managers.[67] In 1956, the Conservation Foundation asked the USDA to cease all use of DDT until its safety could be proven conclusively.[68] The North Carolina game department compared spreading insecticides on fields to exploding an "H-bomb in the pea patch."[69] Wildlife biologists began discussing what they called the "pesticide problem": the chemicals meant to protect the nation—like the atomic bomb—might bring security, but at the same time they undermined America's sense of safety.[70]

Public anxiety about the imported fire ant program grew out of these generalized fears and intensified when the USDA announced that, as part of its eradication program, it would use two pounds of heptachlor per acre, the equivalent of forty pounds of DDT, or eight times the amount needed to kill wildlife.[71] (Why so much if there was no research on the amount of insecticides needed to control the ant? I suspect that it was because, as Waldo Lee Popham said, "nothing less than two pounds is effective" against the white-fringed beetle. The fire ant program needed to be structured to control both insects.)[72] Concerned, conservation organizations asked the USDA to reconsider its plans. As it became clear that the program would proceed, they prepared for the worst.[73] In late 1957 the USFWS assigned wildlife biologist

67. The "boon versus bane" language is from P. F. Springer, "Insecticides: Boon or Bane?" *Audubon* 58 (May–June 1956): 128–30; P. F. Springer, "Insecticides: Boon or Bane?" *Audubon* 58 (July–August 1956): 176–78.

68. D. L. Miller to C. H. Hoffman, 28 June 1957, insecticides, fish and wildlife file, box 64, Entomology Research Division: General Correspondence, 1954–1958, 5 (UD), ARS papers.

69. J. B. DeWitt, *H-Bomb in the Pea Patch* (Raleigh: North Carolina Wildlife Resources Commission, 1957).

70. For the pesticide problem, see J. L. George, "The Pesticide Problem: Wildlife—The Community of Living Things," May–June 1960, folder 221, box 135, Paul B. Sears papers, 663, Manuscripts and Archives, Yale University (hereafter Sears papers); J. L. George, *The Pesticide Problem* (New York: Conservation Foundation, 1957). For fears of the bomb, see P. S. Boyer, *By the Bomb's Early Light: American Thought and Culture at the Dawn of the Atomic Age* (New York: Pantheon Books, 1985).

71. J. B. DeWitt, "Fire Ant Eradication Program," fire ant file, box 22, Wildlife Research Correspondence, 254 (UD), USFWS papers.

72. U.S. Congress, Subcommittee on Fisheries and Wildlife Conservation, Committee on Merchant Marine and Fisheries, *Coordination of Pesticide Programs*, 86-2, 1960, 70.

73. For an example of early conservationist concern, see "Peril in Attack on Fire Ant Seen," *Washington Post*, 18 February 1958, fire ant file, box 30, Entomology Research Division: General Correspondence, 1954–1958, 5 (UD); I. Gabrielson to E. T. Benson, 19 February 1958, regulatory crops-fire ants file, box 944, both in ARS papers.

Walter Rosene to monitor the program. A few months later, the National Audubon Society hired wildlife biologist Harold Peters to do the same. In the summer of 1958, the Conservation Foundation sent a pair wildlife biologists to the South, John George and Robert Rudd. These observers joined Clarence Cottam, a former official with the USFWS who at the time of the eradication campaign headed the Welder Wildlife Foundation in Sinton, Texas, and wildlife biologists associated with the state governments and local universities, all gauging the fire ant program's effect on the South's fauna.

In contrast to the shared unease about insecticides among the conservation community, there was a diversity of opinions about the ant. The work of the Hays brothers convinced the National Audubon Society, the Conservation Foundation, and the Sport Fishing Institute that the ant was not a threat to wildlife.[74] Even Ziebach had by this time come to consider the ant only a minor pest.[75] "In 1948 this column pointed out the Argentine fire ant was damaging to small wildlife, some crops suffered some damage, but the great detrimental effect of the ant was its nuisance value," Ziebach wrote in the *Mobile Press-Register*, retooling history to fit a new reality: an insect that he once had claimed did more damage than the boll weevil was now merely an annoyance.[76] Ralph Allen, a biologist with the Alabama Department of Conservation who had helped to organize the department's involvement in Operation Ant, evinced a similar change of heart. "The fire ant," he wrote, "is a nuisance, but not an economic pest."[77] The National Audubon Society's Harold Peters used the Hays brothers' studies to argue that the ant was "entering into the natural balance."[78] In the early 1960s, the research findings of Auburn graduate student A. S. Johnson seemed to confirm

74. For a description of the National Audubon Society's reaction, see J. Fluno, memo, 11 August 1958, insects affecting man and animals branch fire ant file, box 84, Entomology Research Division: General Correspondence, 1954–1958, 5 (UD), ARS papers. For the Conservation Foundation's acceptance of the work, see "Zoological Society Report Raps at Fire Ant Program," [newspaper unknown,] 15 December 1958, fire ant file, box 198, PPC papers. For the Sport Fishing Institute; see "Fire Ant Eradication Program Condemned," *Sport Fishing Institute Bulletin* 79 (June 1958): 1. For the dissemination of the work of the Hays brothers, see H. S. Peters to D. L. Leedy, 22 July 1958, fire ant file, box 22, Wildlife Research Correspondence, 254 (UD), USFWS papers; J. L. Abbot to Robert H. Michel, 2 August 1961, fire ant file, box 93, PPC papers.

75. For Ziebach's acceptance of the Hays brothers' work, see B. Ziebach, "Some Pertinent Data on Pesky Fire Ants," *Mobile Press-Register*, 10 March 1958, 3A.

76. B. Ziebach, "Fire Ant Problem," *Mobile Press-Register*, 22 June 1958, 6B.

77. R. Allen Jr., "The Fire Ant Eradication Program and its Effect on Fish and Wildlife," fire ant file, SG 17018, ALDoC papers.

78. H. S. Peters, "Report from Research of Pesticides on Wildlife," carton 4, series 1393, R. A. Gray Building, Florida State Archives (hereafter FL papers).

this conclusion. His studies showed that the ant ate only 8 percent of hatching quail chicks, and many of those may have been dead before the ants started eating.[79] (Perhaps changes in the ant's behavior also prompted these revised opinions; as the insect's population declined in some areas it may therefore have been less desperate for food and less interested in eating quail.) Not everyone concurred, though. Ross Leffler, for example, assistant secretary of the interior in charge of the USFWS, said that his agency "wholeheartedly agree[d] that this insect is an unmitigated pest and . . . strongly endorse[d] the objective of its complete elimination," and even Stoddard warned quail farmers that the imported fire ant "MAY well prove to be the greatest threat to the sport of quail shooting that has ever developed."[80]

The variety of opinions about the ant notwithstanding, those who saw the insect as a terrible scourge and those who saw it as a mere nuisance almost unanimously agreed that the use of insecticides posed a far greater threat. Ernest Swift, president of the National Wildlife Federation, compared the eradication program to scalping a patient to stop dandruff: the cure was worse than the disease.[81]

Throughout the winter of 1957 and into the spring of 1958, wildlife enthusiasts fretted over the possible consequences of the eradication program. Then, in April, they saw their worst fears realized. Biologists with the Alabama Department of Conservation surveyed fields in Autauga County, Alabama after the agriculture department sprayed heptachlor. Scanning a hundred-acre plot, they found sixty-eight dead animals, mostly birds, but also some small mammals (fig. 11). William Drinkard, the head of the department, called for an immediate end to the program. "I am now convinced that a large percentage of valuable wildlife, song birds and fish of the state are being destroyed by these treatments," he told a newspaper reporter, perhaps as much as 75 percent of the state's fauna. The study was not conclusive, and not everyone agreed that it could be taken as evidence that the fire ant program was ill conceived. Arant, for example, said, "We know the treatments as

79. A. S. Johnson, "Antagonistic Relationships between Ants and Wildlife with Special Reference to Imported Fire Ants and Bobwhite Quail in the Southeast," *Proceedings of the Annual Conference of the Southeastern Association of Game and Fish Commissioners* 15 (1961): 88–107; A. S. Johnson, "Antagonistic Relationships between Ants and Wildlife" (Master's thesis, Auburn University, 1962).

80. R. L. Leffler to E. T. Benson, 2 December 1957, pesticides, 1949–1960 file, box 24, Wildlife Research Correspondence, 254 (UD), USFWS papers (first quotation); H. L. Stoddard, "Report on 'Imported' Fire Ant Situation on or Near Game Preserves of Thomasville (GA), Tallahassee (FL) Region," 10 March 1958, fire ant file, box 197, PPC papers (second quotation).

81. National Wildlife Federation, press release, 21 November 1957, fire ant file, box 65, Entomology Research Division: General Correspondence, 1954–1958, 5 (UD), ARS papers.

Figure 11. This picture shows some of the dead animals collected in Autauga County, Alabama after the USDA had sprayed for the imported fire ant. The picture was featured in numerous publications and became a symbol that many of the USDA's opponents rallied around, proof that insecticides were a greater danger than the imported fire ant. (From R. H. Allen Jr., "The Fire Ant Eradication Program and its Effect on Fish and Wildlife," fire ant file, SG 17018, Alabama Department of Conservation, administrator's files, 1943–1951, Alabama Department of Archives and History, Montgomery, Ala.)

they are being applied are killing some birds, mammals, and other forms of wildlife; however, we do not know the extent of the kill or whether or not it is sufficient to reduce the overall populations." And Daniel Janzen, heading the USFWS's Bureau of Sports Fisheries and Wildlife (the branch of the agency investigating the fire ant program) said, "Further studies are needed to establish whether the observed losses are atypical or whether they are replicated throughout the range of the imported fire ant."[82] Drinkard, however, insisted, "Seeing them dead in the field following fire ant treatment is strong enough evidence for me."[83] Over the next few months, other investigations confirmed Drinkard's instinct.

82. Notes on telephone conversation with F. S. Arant, 6 May 1958, fire ant file, box 197, PPC papers (first quotation); D. H. Janzen, "Effects of the Fire Ant Eradication Program upon Wildlife," 25 May 1958, fire ant file, box 30, Entomology Research Division: General Correspondence, 1954–1958, 5 (UD), ARS papers (second quotation).

83. "Alabama's Fire Ant Control Program," *Alabama Conservation* 29 (February–March 1958): 4–6; "Insecticide Destroying Wildlife: Fire Ant Control in Alabama Said Proving Dangerous," *Mobile Press-Register,* 20 April 1958, 1. See also Alabama Department of Conservation, *Report for the Fiscal Year October 1, 1957–September 30, 1958,* 90–92.

In Georgia, the USFWS's Walter Rosene found that quail populations on treated land were about 80 percent lower than on untreated land.[84] Wildlife biologists with Auburn, LSU, and the State of Texas substantiated Rosene's findings. Fourteen out of sixteen quail coveys disappeared after treatment of a thirty-six hundred acre tract, Auburn biologists found—and the other two coveys spent significant time on untreated land.[85] Sparrows, meadowlarks, towhees, turkeys, cardinals, thrushes, thrashers, blue jays, mockingbirds, warblers, rails, snipes, juncos, woodpeckers, wrens, rabbits, foxes, rats, mice, fish, frogs, snakes— all died.[86] A year later, bird populations were depressed by about 25 percent; two years later, their numbers remained 35 percent below pretreatment levels.[87] In Louisiana, biologist Leslie Glasgow found thirty-six dead birds three days after the USDA sprayed heptachlor on one 300-acre plot; thirteen dead animals ten days after a 400-acre stretch of land was sprayed; fifty-eight ten days after a 240-acre field was sprayed. Where he could count fifty-four live birds on each acre before the USDA planes flew, he found less than twenty in the days after the insecticide fell from the sky.[88] And in Hardin County, Texas, state wildlife biologist Dan Lay examined a 1,400-acre farm, finding ninety-one dead birds, two nutria, three rabbits, one squirrel, and two raccoons in one month. Fish floated to the surface of ponds. The quail population plunged by 77 percent in ten days and the remaining bobwhite looked frail.[89]

84. W. Rosene, "Whistling-Cock Counts of Bobwhite Quail on Areas Treated with Insecticides and Untreated Areas, Decatur County, Georgia," *Proceedings of the Southeastern Association of Game and Fish Commissioners* 12 (1959): 243.

85. D. W. Speake, "Fire Ant Eradication and Fire Ants in Alabama," insecticides-DDT-wildlife file, box 1, Entomology Research Division: Entomology Director's Correspondence, 1959–1965, 1055 (A1), ARS papers.

86. Ibid.

87. L. A. Parker to S. Bergen, 23 February 1960, fire ant file, box 22, Wildlife Research Correspondence, 254 (UD), RG 22, USFWS papers.

88. L. Glasgow, "Studies on the Effect of the Imported Fire Ant Control Program on Wildlife in Louisiana," *Proceedings of the Southeastern Association of Game and Fish Commissioners* 12 (1959): 250–255; D. H. Janzen, "Effects of the Fire Ant Eradication Program upon Wildlife," 25 May 1958, fire ant file, box 30, Entomology Research Division: General Correspondence, 1954–1958, 5 (UD), ARS papers. See also J. D. Newsom, "A Preliminary Progress Report of Fire Ant Eradication Program Concordia Parish, Louisiana, June, 1958," *Proceedings of the Southeastern Association of Game and Fish Commissioners* 12 (1959): 255–57.

89. D. H. Janzen, "Effects of the Fire Ant Eradication Program upon Wildlife," 25 May 1958, fire ant file, box 30, Entomology Research Division: General Correspondence, 1954–1958, 5 (UD), ARS papers; D. W. Lay, "Fire Ant Eradication and Wildlife," *Proceedings of the Southeastern Association of Game and Fish Commissioners* 12 (1959): 248–50. See also D. W. Lay, "Count Three for Trouble," July 1958; D. W. Lay, "Aftermath of Waste," October 1959, both in folder 581, Rachel Carson papers, YCAL 46, Beinecke Rare Book Library, Yale University (hereafter Carson papers).

Wildlife biologists shipped the dead animals back to the USFWS laboratory in Patuxent, Maryland, where biochemist James DeWitt examined the bodies. DeWitt had worked for the USDA before joining the USFWS and enjoyed respect in both agencies.[90] He found enough heptachlor epoxide (a metabolic by-product of heptachlor) to account for the deaths.[91] He also found evidence that the insecticides climbed the ecological ladder, passing along the chemicals from the insects to the animals that ate them. In Louisiana, for example, earthworms stored heptachlor epoxide in their bodies. Woodcocks could eat a lethal dose in a month. And, if they did not die from the poisons, they could carry them across the country. Louisiana's wetlands hosted many migrating birds that could ingest the chemicals, then fly elsewhere, be eaten, and introduce the poisons to other environments.[92]

By the end of 1958, those who earlier had been reticent about expressing their concerns over the eradication program were now compelled to speak out; outrage began to spread even beyond the conservation community. In November, Auburn wildlife biologist Dan Speake told the Audubon Society, "Such a wide range of vertebrates and invertebrates were affected that the term *zooicide* might well be substituted for *insecticide* in this case."[93] Five months later, Janzen said that the accumulated studies provided "strong evidence that present fire ant control practices entail unacceptable hazards to wildlife resources" and demanded "that there be a cessation of present application methods and dosages of these chemicals until safe procedures can be determined."[94] The Mississippi Wildlife Federation, the Alabama Wildlife Federation, the Georgia Sportsmen's Federation, and Audubon affiliates across the country lobbied the USDA to

90. W. W. Dykstra to Chief, Branch of Research, 3 August 1958; J. L. Buckley to D. H. Janzen, 12 August 1959, both in fire ant file, box 22, Wildlife Research Correspondence, 254 (UD), USFWS papers.

91. D. H. Janzen, "Effects of the Fire Ant Eradication Program upon Wildlife," 25 May 1958, fire ant file, box 30, Entomology Research Division: General Correspondence, 1954–1958, 5 (UD), ARS papers. For a summary of DeWitt's work, see J. B. DeWitt, "Pesticidal Residues in Animal Tissues," *Transactions of the Twenty-fifth North American Wildlife and Natural Resources Conference* (1960): 277–85.

92. The woodcock data is discussed in R. Rudd, *Pesticides and the Living Landscape* (Madison: University of Wisconsin Press, 1964), 264.

93. D. W. Speake, "Fire Ant Eradication and Fire Ants in Alabama," insecticides-DDT-wildlife file, box 1, Entomology Research Division: Entomology Director's Correspondence, 1959–1965, 1055 (A1), ARS papers. Emphasis in original.

94. D. H. Janzen to B. T. Shaw, 29 April 1959, complaints-2-fire ant file, box 211, PPC papers (first quotation); D. H. Janzen to B. T. Shaw, 7 April 1959, regulatory crops, 1959, fire ants file, box 113, General Correspondence, 1954–1966, 1 (UD), ARS papers (second quotation).

call a halt to the program.[95] In 1960, *True,* "the magazine for men"—dedicated to regaling readers with tales of masculine heroism—revisited the pesticide problem. Previously, the magazine had made the case against the imported fire ant and the need for its eradication, and then sneered at the Alabama Department of Conservation for suggesting that insecticides were a worse threat.[96] This time, the ant was not mentioned, but the writer considered insecticides so dangerous that he cut off his shoes and threw them away after he walked through a recently sprayed field. Even he-men were now leery of the chemicals.[97] Summarizing much of the growing opposition to insectides' use, the ornithologist George Wallace called the imported fire ant program "[o]ne of the worst biological blunders that man has ever made."[98]

In its defense, the USDA could point to work by University of Georgia wildlife biologist James Jenkins. His findings showed that even when quail populations plummeted after insecticide was sprayed, their numbers quickly rebounded.[99] Jenkins's results, though, I suspect, were an artifact of his experimental design. He observed only very small areas (one was less than seven acres, another forty acres, in contrast to the thousands of acres sprayed by the USDA elsewhere), so it was easy for quail from off-site to repopulate the sprayed areas. In places where the USDA treated more substantial blocks of land, such repopulation was more difficult. Studies by other wildlife biologists, however—studies painting a bleak picture of the results of insecticides' application—cannot be so easily dismissed. Though only a few studies were formally conducted, everywhere the

95. For Mississippi, see "Mississippi Federation Blasts Fire Ant Control Program," *Conservation News* 23 (1 October 1958): 10–11. For Alabama, see J. L. Abbot to Members of the Board of Directors—Alabama Wildlife Federation, 25 June 1959, box 1, Abbot file, Correspondence of Assistant Secretary for Fish and Wildlife Ross L. Leffler, 790 (A1), Secretary of Interior papers, RG 48, National Archives and Records Administration II, College Park, Md. (hereafter USDI papers). For Georgia, see "Discussions of the Imported Fire Ant at Meetings of Georgia Sportsmen's League," 15–17 October 1959, Peters file, NAS papers. For Audubon affiliates, see L. E. Dickinson to E. T. Benson, 14 April 1959, complaints-2-fire ant file, box 211, PPC papers.

96. D. Mannix, "The Fire Ants Are Coming," *True,* August 1958, 64–67, 96; "Letters to the Editor," *True,* November 1958, 7.

97. H. Stilwell, "Farm Fallout Can Kill You!" *True,* March 1960, 34–36, 76–78.

98. G. Wallace, "Insecticides and Birds," *Audubon* 61 (January–February 1959): 35.

99. For an overview of Jenkins's work, see C. T. Wilson and J. H. Jenkins, "Effects of the Imported Fire Ant Program on Wildlife in Decatur County, Georgia," 10 December 1959, fire ant file, box 93, PPC papers; J. H. Jenkins, "A Review of Five Years' Research on the Effects of the Fire Ant Control Program on Selected Wildlife Populations," *Bulletin of the Georgia Academy of Science* 21 (1963): 3. For the USDA's holding it up as a model, see M. R. Clarkson to D. H. Janzen, 4 June 1958, pesticides 1949–1960 file, box 24, Wildlife Research Correspondence, 254 (UD), USFWS papers.

wildlife biologists looked they found dead wildlife.[100] Wildlife biologists knew that finding even a few dead animals was rare. Usually scavengers gobbled them before any scientist happened by.[101] That they found so many dead animals, and so easily, proved that the insecticides killed significant numbers of wild animals. There is every reason to believe that the insecticides decimated the South's fauna.

The wildlife deaths bothered conservationists on a number of different levels. Some, the Conservation Foundation's Robert Rudd conceded, did not want to see a single bird die; others worried that the deaths might "boomerang" and hurt agriculture: since songbirds ate insects, their absence would lead to renewed explosions of pest populations.[102] Most, though, were not so sentimental that they despaired over every death, nor so naïve that they thought that birds played a significant role controlling insects.[103] Many wildlife biologists, after all, doubted the importance of predation generally. Other opponents of the USDA worried that the imported fire ant program violated the principles of conservation. Clarence Cottam told one agricultural official, "I am not opposed to reasonable controls where there is a proven need. I believe, however, there is a moral obligation to see that other economic and cultural resources are not unnecessarily damaged in the process."[104] It was Cottam's belief that the USDA ignored these other resources. Further, the Sport Fishing Institute noted that the cost of the eradication program "outweigh[ed] the value of the agricultural commodities involved. In dollar value alone, Alabama citizens annually spend $36 million for needed goods and services out [sic] fishing. This is considerably more than the annual value of hogs, dairy products, eggs, forest products, peanuts or cotton seed!"[105] The Conservation Foundation's John George made a similar case: "In order to survive economically," George wrote, "the modern farmer must commit himself to efficient mass production requiring large capital investment, and pesticides are one of his chief

100. D. H. Janzen, "Effects of the Fire Ant Eradication Program upon Wildlife," 25 May 1958, fire ant file, box 30, Entomology Research Division: General Correspondence, 1954–1958, 5 (UD), ARS papers.

101. W. Rosene and D. Lay, "Disappearance and Visibility of Quail Remains," *Journal of Wildlife Management* 27 (1963): 139–42.

102. For Rudd's opinion, see C. C. Fancher to E. D. Burgess, 27 June 1958, imported fire ant-wildlife losses file, box 198, PPC papers. For examples of worries about the implications of bird deaths for agriculture, see "Peril in Attack on Fire Ant Seen," *Washington Post*, 18 February 1958, fire ant file, box 30, Entomology Research Division: General Correspondence, 1954–1958, 5 (UD), ARS papers; C. Kelly to J. Sparkman, 20 March 1959, complaints-2-fire ant file, box 211, PPC papers.

103. For example, R. Clement to H. S. Peters, 26 June 1959, Peters file, NAS papers.

104. C. Cottam to M. Shurtleff, 14 October 1958, fire ant file, box 197, PPC papers.

105. "Fire Ant Eradication Program Condemned," *Sport Fishing Institute Bulletin* 79 (June 1958): 1.

aids. However, chemicals cost money, can kill wild and domestic animals, and cause a toxicity hazard to man. Therefore they should not be used merely as a matter of course, but need for their use should be clearly established. With respect to the imported fire ant pest, this has not been done."[106] The question, George realized, was how to balance competing demands and create the most efficient production system possible. The imported fire ant program, biased in favor of the USDA and agricultural production, was not efficient and so deserved condemnation.

Concerns that the eradication effort undermined the gospel of efficiency led conservationists to a more serious critique of the USDA. Why did the USDA spray so much insecticide to control an ant that might or might not have been a serious pest, but that in any case seemed to pose far less of a threat to the South's animals than the chemicals? Conservationists suggested that something was askew with American public policy. The USDA was, in the phrase of historian Pete Daniel, a "rogue bureaucracy," unbeholden to the American public and unresponsive to its critics.[107] Opponents of the fire ant program offered two possible reasons for the department's cavalier disregard of wildlife. First, the USDA may have been captured by the chemical industry and so launched the fire ant program to line the pockets of its corporate allies. Wildlife advocates had no proof of this contention and, I think, there is little evidence to support it: the USDA decided to eradicate the ant in response to complaints about the insects and because of changes in its bureaucratic structure, not because the chemical industry used political pressure to influence the department's decisions. Still, evidence or no, the USDA's opponents accused it of sacrificing the public good for filthy lucre. Cottam, for example, said of one eradication advocate, "So much of his research funds have come from the pesticide industry, and they have wined and dined him for such a long time, that I am afraid he feels a little beholden to them as his actions and writings indicate. I am afraid that his actions are a clear expression of that old feeling that 'he who pays the piper calls the tune.'"[108] "Boondoggle" was a favored word among the USDA's critics.[109]

Second, the USDA's opponents suggested that the department perverted the gospel of efficiency because it was making a play to increase

---

106. J. L. George, *The Program to Eradicate the Imported Fire Ant* (New York: Conservation Foundation, 1958), 31.

107. P. Daniel, "A Rogue Bureaucracy: The USDA Fire Ant Campaigns of the Late 1950s," *Agricultural History* 64 (1990): 99–114

108. C. Cottam to R. Carson, 26 January 1961, folder 774, Carson papers.

109. See, for example, D. Ferguson, "Fire Ant Eradication—Grandiose Boon-Doggle," 20 March 1969, general information to 1970 file, Series 2012: Fire Ant Correspondence, Mississippi

its bureaucratic power. Fears of bureaucracy have deep roots in U.S. society, but they assumed a dramatic intensity during the Cold War.[110] Americans worried that in building a national security system against communism, the nation would follow in the path of its enemies, since national defense required centralized control and secrecy and conformity—characteristics of totalitarian states.[111] The USDA's eradication program seemed an instantiation of these fears: the department ignored public protest and blanketed the South with poisons without regard to location, land use, or ownership. The South was caught under the "steam-roller of bureaucracy," Ziebach said. "As long as Congress puts millions of dollars into the hands of incompetents with no better solutions than insecticides as a control, Congress will find it is only building up bureaus of state and federal jobs—and courting biological disaster."[112] This became a common refrain. Another critic, for example, complained, "A mass spray program violates our rights of no trespass without our express consent. This is just one more step in the direction we are going whereby our government, instead of existing to serve the citizens, is fast becoming our master without our consent. Do we have to wait until we have been a dictatorship before we recognize this trend and call a halt to it?"[113] Resist the USDA, Clarence Cottam said, and resist becoming "mere numbered pawns of the State."[114]

The power of this critique can be seen in its influence on J. Lloyd Abbot. A Mobile nurseryman and former naval officer, Abbot had been a dogged defender of attempts to control the white-fringed beetle.[115] When he had heard that the USDA planned to eradicate the insect, he was ecstatic. "That is splendid," he said. "It is hoped that you will indoctrinate the personnel . . . who are working on the white-fringed beetle operations, down to the last man, on that type of thinking and

---

Department of Agriculture and Commerce papers, Mississippi Department of Archives and History, Jackson, Miss.; "Latest Boondoggle Proposed by the Agri-Chemical Complex," *Audubon* 72 (1970): 143; I. R. Barnes, "The $20,000,000 Fire Ant Boondoggle," correspondence, internal file, box 71, EDF papers.

110. On American antigovernment sentiments generally, see G. Wills, *A Necessary Evil: A History of American Distrust of Government* (New York: Simon and Schuster, 1999).

111. For these fears, see M. Hogan, *A Cross of Iron: Harry S. Truman and the Origins of the National Security State, 1945–1954* (New York: Cambridge University Press, 1998).

112. B. Ziebach, "Fire Ant Problem," *Mobile Press-Register*, 22 June 1958, 6B.

113. "Experiments of USDA Violate Individual Rights," n.d., fire ant file, box 94, PPC papers.

114. "The Pesticide Problem," carton 4, series 1393, FL papers.

115. Abbot's biographical data is from J. L. Abbot Jr. to author, 23 November 1999. His view of the beetle is from J. L. Abbot to E. T. Benson, 22 February 1954, white-fringed beetle correspondence with J. L. Abbot file, box 29, PPC papers.

determination, and also tell them that if you hear of any one of them ever again stopping some person from using the word eradication, that you will have him up on the carpet for a complete explanation of why he has such a negative and defeatist approach to his responsibilities."[116] But, then, when southern states proved resistant to the eradication program, the USDA seemed to drop its interest in the beetle and turn all of its attention to the ant. Abbot soon heard about the work of the Hays brothers and was confused. Why suddenly attack an insect that was not even a pest and ignore one that was supposed to be a terrible scourge?[117] Abbot suspected that the USDA did not care so much about eradicating the ant as it did about increasing its bureaucratic power and winning public acclaim by distributing free or subsidized insecticide.[118] Here was the seed of the American garrison state. "The threat to the continued existence of our democracy," he wrote, "and whether or not we are going to be taken over by *internal* bureaucracies, could not be more clearly illustrated than it is by this whole reprehensible situation."[119] Abbot joined the Alabama Wildlife Federation to coordinate opposition to the program.

The imported fire ant eradication program thus presented a dilemma for conservationists. On the one hand, many loathed the insect, despite the work of the Hays brothers. On the other hand, the insecticides used to kill it also seemed a threat, as did the bureaucracy that spread the chemicals so cavalierly. "Mention the word toxicant to many outdoorsmen and you get an unpleasant reaction. Mention fire ants and you get another disdainful look," wrote Texas outdoor columnist John Thompson in 1962. "Most area hunters have witnessed or have heard unfavorable things about both subjects. There has been widespread controversy over the use of toxicants and insecticides the past few years. There has also been major concern over the spread of the imported fire ants. Obviously, here are two evils which outdoorsmen dislike."[120] In the face of the destruction wrought by the insecticides, most (although not all) conservationists decided that the

---

116. J. L. Abbot to L. F. Curl, 1 September 1955, white-fringed beetle correspondence with J. L. Abbot file, box 29, PPC papers.

117. J. L. Abbot to B. T. Shaw, 25 October 1957, fire ant files (1957), PPC papers.

118. Abbot wrote a huge number of letters, but for a summary of his views, see J. L. Abbot to R. H. Michel, 2 August 1961, fire ant file, box 93, PPC papers.

119. J. L. Abbot to E. V. Smith, 24 October 1958, box 2, Draughon papers. Emphasis in original.

120. J. Thompson, "Insecticide for Fire Ants," 11 October 1962, fire ant file, box 128, PPC papers.

insecticide was the greater of the two evils.[121] Some supported this contention by noting the research of the South's university entomologists, but even without these claims they had plenty of evidence that the insecticides killed large numbers of wildlife—evidence, they believed, that also showed that public policy was off kilter and that the USDA was out of control. The imported fire ant eradication program was not a testament to the gospel of efficiency, the USDA's opponents said, but a way for the federal department to seize power. The ant, by contrast, seemed not so savage. As J. Lloyd Abbot wrote, "The fire ant is no boll weevil or white-fringed beetle. Perhaps state and federal aid is the only answer to a nuisance which observes no property lines and even makes its home along highway and railroad rights of way. But this is not a Martian invasion, nor the Year of the Locust."[122]

The Imported Fire Ant and the Emergence of the Environmental Movement

Robert Rudd, Harold Peters, and, especially, Rachel Carson offered a different critique of the imported fire ant program. Along with many wildlife biologists, they held that the fire ant was not as bad as advertised and that the insecticides were worse, but from this they drew a different conclusion. The problem was not that the USDA had perverted the gospel of efficiency; indeed, the imported fire ant program epitomized the conservationist's ideal—and revealed the inadequacies of that philosophy. The problem was that the USDA only saw nature as a collection of resources. But nature was more than that, Rudd, Peters, and Carson argued, more than resources that could be controlled and molded into any shape that humans saw fit. Nature was an intricate web of relationships that sustained human life. Insecticides and other chemicals could pass through this network and, carried by food and water, come to lodge in the human body, where they might cause cancer and other diseases. In computing the gospel of efficiency, proponents of the conservation philosophy ignored these repercussions and the missing variables invalidated their calculus.

This critique represented the leading edge of the emerging environmental movement. According to historian Samuel Hays, environmentalism emerged in the second half of the twentieth century not as part

---

121. For examples of wildlife enthusiasts who continued to see the ant as the greater threat, see J. D. Donehue, "No Conflicts Here in Fire Ant Fight," 14 September 1958, fire ant file, box 197; F. Miller, "Fishing with Floyd," 22 June 1961, fire ant files (1961), both in PPC papers.

122. J. L. Abbot, "Elements of Myth in the Fire Ant Peril," *Montgomery Advertiser*, 11 March 1957, fire ant-news clippings file, box 65, Entomology Research Division: General Correspondence, 1954–1958, 5 (UD), ARS papers.

of the youth movement or counterculture revolution—although it became associated with both of these—but as a reflection of broader and more fundamental social changes. As Americans became better educated and wealthier, they sought—in addition to necessities (food, water, shelter) and conveniences (refrigerators, radios, and washing machines, among other things)—what Hays calls "amenities": the qualities of a good life. Americans wanted clean water to drink, clean air to breathe, and natural areas in which to play. Thus, environmentalism stemmed "from a desire to improve personal, family, and community life. . . . An interest in the environmental quality of life is to be understood simply as an integral part of the drives inherent in persistent human aspiration and achievement."[123] Environmentalism also represented, according to historian Thomas Dunlap, the emerging consensus that nature was not an isolated collection of resources, but a complexly integrated ecosystem of which humans were a part. Altering nature's balance could have profound consequences for human beings.[124] Aldo Leopold, shedding his conservationist ideals to become the seminal philosopher of the environmental movement, expressed the new ideal in his posthumously published *A Sand County Almanac.* "A thing is right," Leopold wrote, "when it tends to preserve the integrity, the stability, and beauty of the biotic community. It is wrong when it tends otherwise."[125] The "good life" depended on preserving elements of nature—even seemingly noxious parts, such as predators—not just using them, for only by preserving nature could Americans be guaranteed the amenities that they wanted. Judged by this standard, the fire ant program did not protect the American way of life, but threatened it, killing wildlife and contaminating the environment on which the nation's quality of life depended.

The distinction between conservation and environmentalism is clear, but it can be overdrawn, as the evolution of Leopold's thought shows: the environmental movement grew, in part, out of conservationist belief.[126] In the late 1950s, as the controversy over how to deal

123. S. P. Hays, *Beauty, Health, and Permanence: Environmental Politics in the United States, 1955–1985* (New York: Cambridge University Press, 1987), 5.

124. T. R. Dunlap, *Saving America's Wildlife: Ecology and the American Mind, 1850–1990* (Princeton: Princeton University Press, 1988), 98–105.

125. A. Leopold, *A Sand County Almanac* (New York: Ballantine Books, [1949] 1970), 262. On the evolution of Leopold's thought see S. Flader, *Thinking Like a Mountain: Aldo Leopold and the Evolution of an Ecological Attitude Toward Deer, Wolves, and Forests* (Columbia: University of Missouri Press, 1974).

126. On this point, see A. Rome, *The Bulldozer in the Countryside: Suburban Sprawl and the Rise of American Environmentalism* (New York: Cambridge University Press, 2001), 8; J. B. Buhs, "Dead Cows on a Georgia Field: Mapping the Cultural Landscape of the Post–World War II American Pesticide Controversies," *Environmental History* 7 (2002): 112.

with the imported fire ants warmed, both Robert Rudd's Conservation Foundation and Harold Peters's Audubon Society were struggling to accommodate the rise of environmentalism. The Audubon Society's historian notes, "Concerns about personal health and the quality of their lives were attracting a vastly wider public to the movement, including people who did not know the difference between a bluebird and a blue jay but who felt that the world was somehow out of control and were looking for innovative solutions. Like penitents flocking to a church in times of pestilence, the newcomers were welcome, but graybeards among the faithful realized that the congregation would inevitably undergo drastic changes in tone and leadership."[127] Some in the Conservation Foundation also groused about the Audubon Society's new direction, while others embraced it. Foundation member Paul Sears, for instance, said that the society was the preeminent defender of "intangible values": "beauty, leisure, enjoyment of nature, and our fellow beings . . . those things which money cannot measure," but for which "cultures and nations are chiefly esteemed."[128]

Rudd and Peters pushed their respective organizations to embrace the emerging environmental ethic. While John George bemoaned the USDA's failure to properly make production efficient, Rudd attacked production itself, calling it a "fetish," a philosophy as "antiquated as the ox-drawn cart."[129] To produce more crops, American farmers, at the behest of their scientific advisors, simplified ecosystems, replacing complex, diverse communities of plants and animals with field after field of the same crop. It was these simplified ecosystems that turned insects into pests, their population explosions fueled by the bonanza of food available to them.[130] (Rudd did not mention the imported fire ant by name, but it well illustrated his point.) In response to these insect irruptions, he lamented, many entomologists thought only to spray insecticides.[131] But the chemicals did little for the economy—and they hurt humans. America

127. F. Graham Jr., *The Audubon Ark: A History of the National Audubon Society* (New York: Knopf, 1990), 228.

128. On Conservation Foundation members' objection to the National Audubon Society's reorientation, see A. C. Worrell to P. B. Sears, 28 August 1960, folder 290, Sears papers. Sears's comments are from P. B. Sears, "The Road Ahead in Conservation," *Audubon* 58 (March–April 1956): 58–59, 80. For a history of the Conservation Foundation, see R. Gottlieb, *Forcing the Spring: The Transformation of the Environmental Movement* (Washington, D.C.: Island Press, 1993), 38–39.

129. R. Rudd, "The Indirect Effects of Chemicals in Nature," file 583, Carson papers.

130. R. Rudd, *Pesticides and the Living Landscape* (Madison: University of Wisconsin Press, 1964), 186–219.

131. Ibid., 184, 286.

already produced too much food, Rudd noted—the federal government had to subsidize farmers to make the industry profitable since overproduction kept prices low—and the pesticides used to make it even more productive threatened human health and killed wildlife.[132] The production fetish was a runaway system, scientists scrambling to increase efficiency without regard to costs or necessity. It was, he told the Audubon Society, a "false god to which are sacrificed a host of other values . . . from which spring the ingredients of the full life."[133] The imported fire ant eradication program epitomized the deficiencies of the gospel of efficiency as a guide to public policy. It "lacked imagination, depth of prior knowledge, and the promise of success that both inspire," Rudd wrote in his 1964 book *Pesticides and the Living Landscape*.[134]

Peters pushed the Audubon Society in another direction, not so much castigating agricultural scientists for obeisance to the production fetish as for ignoring the dangers that insecticides posed to human health. Others had expressed similar worries (Clarence Cottam, for example), but among those who opposed the fire ant program, Peters pushed the argument the farthest until Rachel Carson published *Silent Spring* in 1962.[135] Concern over the health effects of insecticides had received some airing in 1958 during the Gypsy moth trial—when the Mayo Clinic's Malcolm Hargraves linked them to numerous diseases—and animated much of the discussion in the Audubon Society. Peters's boss, society president John Baker, for example, said that chemical insecticides were the greatest threat to life on earth, worse than communism, worse than radioactive fallout.[136] He worried that they could pass from the environment into human bodies, mutating genetic material, their effects resounding for generations.[137] Similar fears surrounded chemical fallout: one study, for example, showed that the radioactive particle Strontium-90, released from the explosion of atomic bombs, drifted to the ground, where it was consumed by cows, contaminating their milk, then passing to humans, where it lodged in the teeth of

132. R. Rudd, "The Indirect Effects of Chemicals in Nature," file 583, Carson papers.

133. Ibid.

134. R. Rudd, *Pesticides and the Living Landscape* (Madison: University of Wisconsin Press, 1964), 36.

135. For Cottam's views, see C. Cottam, "The Uncontrolled Use of Pesticides in the Southeast," *Proceedings of the Southeastern Association of Game and Fish Commissioners* 13 (1959): 17.

136. J. H. Baker, "The Greatest Threat to Life on Earth," *Outdoor America* 23 (June 1958): 4–5.

137. J. H. Baker to R. L. Leffler, 21 January 1958, pesticides 1949–1960 file, box 24, Wildlife Research Correspondence, 1934–1966, 254 (UD), USFWS papers.

children.[138] The interspecies spread of the dangerous chemicals proved to many that nature was an integrated matrix of ecological relationships that needed to be protected to protect human life, justifying the extension of the Audubon Society's interests: the same chemicals that killed birds also killed humans, and the distance between dangers to the environment and dangers to humans was a narrow one. Baker charged Peters with finding "good evidence" illustrating the connection.[139]

Peters thought that he did. In the summer of 1958, Dan Lay told him that several cattle in Texas died after the USDA sprayed for the imported fire ant.[140] And he learned from Otis L. Poitevint, a veterinarian in Bainbridge, Georgia, that the fire ant program had also killed a number of cows in the area. Especially troubling was the case of a two-month-old calf; it had been born months after the USDA last sprayed, indicating that it either was poisoned *in utero* or from its mother's milk—evidence that the insecticides used to kill fire ants could work their way through the web of nature and reach humans who drank milk, just like Strontium-90.[141] Two USDA veterinarians disagreed with Poitevint's and Lay's diagnoses, arguing that some cows had no detectable amounts of insecticide in their bodies and others had too little to account for the deaths.[142] (Frank Arant agreed that the amount of insecticide used in the fire ant program could not kill cattle.)[143] Peters, though, was confident that he had established a link between insecticides and human health. He told one reporter, "Cattle absorb the chemicals from eating grass which has been dusted with the insecticides and, in turn, pass the chemicals to humans through milk and meat." Drawing on the work of the Mayo Clinic's Malcolm Hargraves, Peters noted that once the chem-

138. For a discussion of this study and how fears of atomic fallout interweaved with trepidations about insecticides, see R. H. Lutts, "Chemical Fallout: Rachel Carson's *Silent Spring,* Radioactive Fallout and the Environmental Movement," *Environmental Review* 9 (1985): 214–25.

139. J. H. Baker to H. S. Peters, 27 March 1959, Peters file, NAS papers.

140. The connection between Lay and Peters is noted in C. C. Fancher to supervisors in charge, southern plant pest control region, 26 September 1958, insecticide-2-toxic effects file, box 134, PPC.

141. O. L. Poitevint to R. E. Tyner, 13 October 1959; "Discussions of the Imported Fire Ant at Meetings of Georgia Sportsmen's League," 15–17 October 1959, Peters file, both in NAS papers.

142. R. D. Radeleff, "Twelve Points Agreed upon by All Investigators of Cattle and Other Livestock Losses at Bainbridge, Georgia," 22 April 1958; W. Buck to J. L. Massey, 17 September 1958, both in insecticide-2-toxic effects file, box 134; C. C. Fancher to Southern Plant Pest Control Supervisors and State Plant Pest Control Officials, 6 October 1958; L. J. Padget to E. D. Burgess, 30 October 1958, both in fire ant file, box 197, all in PPC papers.

143. F. S. Arant to W. O. Owen, 10 April 1958, fire ant file, box 197, PPC papers.

icals entered the body, they caused aplastic anemia, lymphoma, and other blood diseases.[144] "This insidious danger," he wrote, "is a threat to people eating vegetables, fruit, and meat grown on chemically treated areas as well as through drinking milk or water from treated areas."[145]

Despite his certainty that he had found the solid evidence Baker wanted, Peters never felt that he received the support from his superiors that he deserved. The National Audubon Society was venturing into new territories and, unsure of its footing, wanted irrefutable evidence to support its claims about the dangers of insecticides. Peters never found sufficient evidence to wholly disprove the claims by the USDA's veterinarians, and Baker was unwilling to risk his society's reputation on disputed findings.[146] Although Peters published his conclusions in the society's magazine *Audubon,* he complained that its editors watered down his reports, and that no one promoted his work.[147] Rudd, too, voiced opinions not always accepted by others in the Conservation Foundation, and seemed to have similar difficulty attracting attention.[148] He spoke at the Audubon Society's national convention and published two articles in *The Nation;* but the book that he wrote for the Conservation Foundation took years to publish—held up by advocates of chemical insecticides who thought that it was biased—and when it finally appeared it had virtually no effect on public thought.[149] The environmentalists' message—that the imported fire ant program endangered the beauty of wildlife, the permanence of nature's balance, the health of humans—reached the public only sporadically, traveled through restricted channels, and did not energize a movement.

144. "Ant Poison Causes Increase in Diseases—Peters," [newspaper unknown,] 29 August 1959, Peters file, NAS papers.

145. "The Hazards of Broadcasting Toxic Pesticides as Illustrated by Experience with the Imported Fire Ant Control Program," insecticides-DDT-wildlife file, box 1, Entomology Research Division: Entomology Director's Correspondence, 1959–1965, 1055 (A1), ARS papers.

146. J. H. Baker to H. S. Peters, 6 April 1959, Peters file, NAS papers.

147. For his complaints, see H. S. Peters to J. H. Baker, 13 October 1958; H. S. Peters to R. Clement, 30 July 1962, both in Peters file, NAS papers. For more on this point, see J. B. Buhs, "Dead Cows on a Georgia Field: Mapping the Cultural Landscape of the Post–World War II American Pesticide Controversies," *Environmental History* 7 (2002): 110.

148. For Conservation Foundation opinions on pesticides different from Rudd's, see A. C. Worrell, "Pests, Pesticides, and People," *American Forests* 66 (1960): 39–81.

149. R. Rudd, "The Irresponsible Poisoners," *Nation* (30 May 1959): 496–97; R. Rudd, "Pesticides: The Real Peril," *Nation* (28 November 1959), 399–401. Rudd's trouble publishing his book is from his discussion of his career, 21 and 22 August 1999, on audiotape, in author's possession. See also R. Van den Bosch, *The Pesticide Conspiracy* (New York: Doubleday, 1978), 62.

Rachel Carson and the Imported Fire Ant

The situation changed in 1962 with the publication of Rachel Carson's *Silent Spring*. Carson built on the work of the South's university entomologists as well as the arguments of wildlife biologists, joining them with her own research about the health effects of insecticides to produce a stinging indictment of the USDA, the chemicals it used, and its eradication ideal. Her book became a bestseller and sparked a national debate over the use of insecticides and the gospel of efficiency that it served.[150] "The 'control of nature,'" she wrote, "is a phrase conceived in arrogance, born of the Neanderthal age of biology and philosophy, when it was supposed that nature exists for the convenience of man. The concepts and practices of applied entomology for the most part date from that Stone Age of science. It is our alarming misfortune that so primitive a science has armed itself with the most modern and terrible weapons, and that in turning them against the insects it has also turned them against the earth."[151] The imported fire ant program, she decided, would be "exhibit A" in her case against the USDA and the eradication ideal, an "outstanding example of an ill-conceived, badly executed, and thoroughly detrimental experiment in the mass control of insects."[152]

Since childhood, Carson had associated nature with living things, a habit of thought reinforced when she studied for a master's degree in biology from Johns Hopkins University and while she was a scientific writer at the USFWS (where she sometimes worked closely with Cottam, until she quit in 1952 to be a full-time author).[153] Carson believed that over the course of millennia biological entities had followed a path appointed by God, slowly adapting to one another and the environment, building ecological relationships in which every organism played an important role.[154] She dismissed the belief of many wildlife biologists that predators were superfluous. In 1961, for example, as Carson was researching *Silent Spring*, Paul Errington, Aldo Leopold's student and the nation's leading expert on predators during the mid-twentieth century, told her that predation was "incidental

150. On the reception of Carson's work and the issues at play, see F. Graham Jr., *Since Silent Spring* (Boston: Houghton Mifflin, 1970).

151. R. Carson, *Silent Spring* (Boston: Houghton Mifflin, [1962] 1994), 297.

152. L. Lear, *Rachel Carson: Witness for Nature* (New York: Henry Holt, 1997), 340 (first quotation); R. Carson, *Silent Spring* (Boston: Houghton Mifflin, [1962] 1994), 162 (second quotation).

153. For Carson's life, see L. Lear, *Rachel Carson: Witness for Nature* (New York: Henry Holt, 1997).

154. *Always, Rachel: The Letters of Rachel Carson and Dorothy Freeman, 1952–1964*, ed. M. Freeman (Boston: Beacon Press, 1995), 249.

in nature . . . It seems to have very little if any actual influence on population levels."[155] Carson, though, devoted a chapter of her book to the relationship between predator and prey and the "dire results of upsetting nature's own arrangements."[156] For her, predators were a vital part of functioning ecosystems.

Humans, though, and the chemicals that they made, stood outside of nature's order. "The chemicals to which life is asked to make its adjustment are no longer merely the calcium and silica and copper and all the rest of the minerals washed out of the rocks and carried in rivers to the sea; they are the synthetic creations of man's inventive mind, brewed in his laboratories, and having no counterparts in nature," she wrote.[157] One finds Carson's preference for the biological over the chemical in her celebration of Edward Fred Knipling's screwworm program (see chapter 2). She called it "fascinating," "brilliant," "spectacular."[158] Knipling, though, as much as the PPC entomologists, aimed to control nature: he wanted to drive an animal to extinction by modifying its genetic makeup.[159] I think Carson overlooked this inconsistency because Knipling's work was rooted in biology, not chemistry, and that made all the difference. Living things belonged to nature, Carson thought. Chemicals did not.

Viewing the controversy over the imported fire ant eradication campaign through this lens resolved ambiguities. The insecticides, Carson argued, destroyed everything that they touched. In *Silent Spring* she recounted the tales of animal deaths collected by Lay and Cottam and Peters and all the rest of the wildlife biologists. She also wrote about the dead two-month-old calf.[160] While researching the book, Carson learned that USDA veterinarians argued that the animals had not been poisoned, but she ignored these claims, satisfied that the animal's death at least suggested that chemical insecticides could pass through the web of nature, lodge in human bodies, and cause disease and death.[161] She spent a chapter documenting the baleful health problems caused by insecticides, dwelling on the possibility that they

155. R. Carson to P. Errington, 6 October 1961, folder 784 and undated response, Carson papers.

156. R. Carson, *Silent Spring* (Boston: Houghton Mifflin, [1962] 1994), 248,

157. Ibid., 7.

158. Ibid., 279–82, 297.

159. See, for example, C. G. Scruggs, *The Peaceful Atom and the Deadly Fly* (Austin: Jenkins Publishing, 1975), 297.

160. R. Carson, *Silent Spring* (Boston: Houghton Mifflin, [1962] 1994), 7.

161. For Carson's contact with the USDA's veterinarians, see R. Carson to R. D. Radeleff, 20 May 1959; R. D. Radeleff to R. Carson, n.d., both in file 833, Carson papers. Carson probably read Poitevint's report in C. Cottam, "Pesticides and Wildlife," file 581, Carson papers.

caused cancer. Only an "autocratic" government, she said, would unleash such a terrifying weapon with such indifference for the repercussions.[162]

Because the ant, on the contrary, was a living thing, Carson asserted it had a place in a balanced natural order—a view that required the suppression of contrary facts as much as the one espoused by the PPC entomologists. E. O. Wilson, for example, told her, "The fire ant is a pest and in past years at least has been a serious one at times." Those who opposed the eradication campaign only "harmed their argument" by denying the ant's depredatory ways, he said. Carson, though, paid no heed to Wilson's admonitions, never even citing his work in *Silent Spring*.[163] She also ignored Ziebach's description of the ant as a plague worse than the boll weevil and Wilson's claim that the red ant was a mutant of the insect that Henry Walter Bates had seen attacking Aveyros in the Amazon basin. By her reckoning, there was no distinction between the red and black ants. The same insect had resided in the South for forty years without incident, Carson wrote, overlooking the furor caused by the ant in the late 1940s.[164] Over time, it had joined the region's ecological web. Carson noted, for example, that the insect ate sugar cane borers, pointing to the research by the entomologists from LSU. When the USDA spread insecticides to kill the ant, the sugar cane borer population exploded—proof that the ant played an important role in the ecosystem.[165] Moreover, its metal-twisting nests were better seen not as a bane for tractor drivers, but as an illustration of the ant's moral virtue as sound steward of the earth. She wrote, "Their mound-building activities serve a useful purpose in aerating and draining the soil."[166] The USDA had created the ant's image as a serious pest out of whole cloth, Carson implied, to help the chemical industry sell insecticides.[167] Now, she was offering the ant's true story. She told J. Lloyd Abbot that *Silent Spring* "thoroughly documented that the fire ant has never been a menace to agriculture and that the facts concerning it have been completely misrepresented."[168]

---

162. R. Carson, *Silent Spring* (Boston: Houghton Mifflin, [1962] 1994), 127.

163. E. O. Wilson to R. Carson, 23 October 1958; E. O. Wilson to R. Carson, 14 May 1959, both in folder 841, Carson papers.

164. R. Carson, *Silent Spring* (Boston: Houghton Mifflin, [1962] 1994), 162–63.

165. Ibid., 255.

166. Ibid., 163.

167. Ibid., 162–63.

168. R. Carson to J. L. Abbot, 6 October 1961, folder 1586, Carson papers.

The Imported Fire Ant: From Savage to Invincible

Silent Spring expressed a vision of the imported fire ant diametrically opposed to the one offered by the USDA. But the ant's rehabilitation was not yet complete. Its name still proclaimed its savagery: the insect was *Solenopsis saevissima,* a mutated form of the ant that Bates had seen attack Aveyros a century ago. Less than a decade after *Silent Spring* appeared in bookstores, though, that was no longer the case. In the late 1960s, three myrmecologists skeptical of E. O. Wilson's science revisited the imported fire ant's taxonomy. In the process, William Franklin Buren, Murray Sheldon Blum, and William Steel Creighton helped to fill in the image of the ant as sketched in *Silent Spring:* the insect did not deserve our hate, but was worthy of admiration. This was an image that emerged out of the interaction between the insect, in all its variability, and people whose intellectual commitments were influenced by the position that they occupied in American society.

The taxonomic status of the imported fire ant became a topic of discussion in 1967, when Creighton met Blum. The two men became fast friends, sharing a sharp sense of humor and deep passion for ants. In the course of their discussions, it was impossible that E. O. Wilson's name not come up: Wilson was his generation's leading myrmecologist, following his taxonomic studies with investigations into the evolution of Melanesian ants and the biochemistry of ant communication. Creighton, though, harbored serious reservations about Wilson's taxonomic work. During the 1950s, when Wilson and William Brown had been establishing themselves within the myrmecological community, Creighton had fought a series of bruising battles with them, the older man eventually deciding that he could not trust Wilson and Brown's taxonomy.[169] Too often, Creighton told Blum, Wilson put theory before facts and, consequently, often got his facts wrong.[170] (By his own admission, Wilson enjoyed floating ideas on the basis of evidence that was more suggestive than definitive.)[171] "If there had been a skating on thin ice contest at the recent Olympics," Creighton said, "Ed would have brought back the gold, silver and bronze medals to the United States."[172] In particular, Creighton told Blum, he doubted that Wilson's studies of fire ant taxonomy merited continued respect. The red fire ant

---

169. For this history, see J. B. Buhs, "Building on Bedrock: William Steel Creighton and the Reformation of Ant Systematics," *Journal of the History of Biology* 33 (2000): 53–64.

170. W. S. Creighton to M. S. Blum, 14 May 1968; W. S. Creighton to M. S. Blum, 22 January 1969, both in Creighton papers.

171. M. Ruse, *Mystery of Mysteries: Is Evolution a Social Construction?* (Cambridge: Harvard University Press, 1999), 184–91.

172. W. S. Creighton to M. S. Blum, 2 March 1972, Creighton papers.

was probably not a mutant, but, rather, the red and black insects belonged to different species.[173] Blum, a biochemist by training, had always thought that some of Wilson's conclusions about ant chemistry were wrong or superficial, but had assumed his taxonomy was solid.[174]

Blum sensed an opportunity. He had become interested in the fire ant after the insect stung his daughter, and now was receiving shipments of fire ants from the PPC for his biochemistry studies.[175] Blum thought that he might redirect this material to taxonomic work, finally checking Wilson's ideas about the fire ants. "Since [Wilson] has 'muddied up the waters,' and many people consider his pronouncements as being tantamount to a message from above, it is obvious that someone who has not been hypnotized will be required in order to settle this issue," he said.[176] Not primarily a taxonomist, Blum wanted Creighton to work with him on the study, but Creighton begged off.[177] He wrote, "Macarthur was the only man that I know that could carry on a major campaign after he was sixty-five. I have fought enough with Wilson and Brown and I want no more of it. Neither, I suspect, do they. However, there is no reason why you cannot get out a paper . . . and, if you do, I will give it my full support."[178]

Rebuffed but not discouraged, Blum turned to William Buren. Buren nursed his own grievances against Wilson. He thought that Wilson and Brown had usurped some of his ideas and published them without sufficient credit.[179] And he worried that Wilson's rapid progress came at the expense of incomplete taxonomic studies. "*Someone* has to do the niddy-griddies to check out Wilson's theories while he continues onward and upward to still greater and greater glories," he said.[180] When Blum broached the subject of revising fire ant taxonomy, Buren jumped at the chance. He had spent a decade working for the Public Health Service and now wanted to return to academia. An authoritative taxonomy of the imported insect would bring him to the attention of universities and, as well, correct Wilson's taxonomic studies. Creighton approved, telling Buren, "It's high time somebody showed up Ed and I hope you and Murray will do it."[181]

173. W. S. Creighton to M. S. Blum, 22 April 1968, Creighton papers.

174. M. S. Blum to W. S. Creighton, 7 August 1967, 25 April 1968, both in Creighton papers.

175. On Blum's daughter, see "Milking Fire Ants," *Newsweek* 76 (9 November 1970): 70.

176. M. S. Blum to W. S. Creighton, 25 April 1968, Creighton papers.

177. M. S. Blum to W. S. Creighton, 7 August 1967, Creighton papers.

178. W. S. Creighton to M. S. Blum, 22 April 1969, Creighton papers.

179. J. B. Buhs, "Building on Bedrock: William Steel Creighton and the Reformation of Ant Systematics," *Journal of the History of Biology* 33 (2000): 63.

180. W. F. Buren to W. S. Creighton, 24 April 1972, Creighton papers. Emphasis in original.

181. W. S. Creighton to W. F. Buren, 19 January 1969, Creighton papers.

Having agreed to work together, Blum and Buren began harvesting fire ants. They needed a cache of black fire ants to compare to the red ones, they needed insects from South America, and they needed a historical collection to compare. No one knew if the black ant even still existed in the United States or if the red one had driven it to extinction, so Blum contacted Leyburn Lewis and H. B. Green for hints about where the black ant might still be surviving.[182] He also traveled to South America to collect the insect in its homeland.[183] Buren combed through museum collections and headed into the field.[184] Eventually, the myrmecologists found populations of the black fire ant living in northeast Mississippi and northwest Alabama.[185] Buren had trouble collecting the ant at first because government entomologists kept killing the insect before he could get to it.[186] He also discovered that he was allergic to its sting and his students refused to allow him near the insect.[187] These colleagues, though, were able to make a healthy collection. Not allergic to the red fire ant, Buren gathered it unhindered.[188]

Sitting down to study ants collected from over five hundred nests, Buren found that, contrary to Wilson's earlier conclusions, he could separate the imported fire ant from *Solenopsis saevissima:* the ant invading North America was not the same ant that had attacked Aveyros. Differentiating between the red and black fire ants was more difficult, but, in addition to color, Buren found ten "subtle" distinctions.[189] The traits so aptly characterized the ant that even Wilson had to admit that

182. H. B. Green to M. S. Blum, 29 April 1968, Creighton papers; L. F. Lewis to M. S. Blum, 26 July 1968, Blum papers.

183. On Blum's work, see W. F. Buren to W. S. Creighton, 15 January 1969; M. S. Blum to W. S. Creighton, 26 January 1969, both in Creighton papers; *Quarterly Report for Cooperative Agreement between University of Georgia and United States Department of Agriculture,* April 1, 1969–30 June 1969, Georgia biology and ecology of the imported fire ants file, box 7, Entomology Division Central Files, 1934–1972, 1061 (A1), ARS papers.

184. W. F. Buren to W. S. Creighton, 15 January 1969, Creighton papers.

185. Ibid.

186. On the killing of the black ant, see ibid.

187. For Buren's allergic reaction, see "Public Hearing to Determine Whether or Not the Registration of Mirex Should be Cancelled or Amended," F.I.F.R.A. Docket no. 293, transcript pages 3817, 3867–68, 14 March 1974, Buren 3/14/74 file, box 62, EDF papers. See also R. B. Rhoades, *Medical Aspects of the Imported Fire Ant* (Gainesville: University Presses of Florida, 1977), 59–60.

188. "Public Hearing to Determine Whether or Not the Registration of Mirex Should be Cancelled or Amended," F.I.F.R.A. Docket no. 293, transcript pages 3817, 3867–68, 14 March 1974, Buren 3/14/74 file, box 62, EDF papers.

189. Quotation from W. F. Buren to W. S. Creighton, 29 January 1969, Creighton papers. See also W. F. Buren to W. S. Creighton, 7 April 1970, Creighton papers, in which he lists the ten characters.

Buren's conclusions were correct.[190] In addition, Blum found differences in the biochemistry of the red and black ants.[191] Creighton crowed in delight. The USDA, he said, now had "some accurate field observations to go on and won't be confused by theoretical exordiums of 'evolution in progress.'"[192] A few federal entomologists attacked Buren's research (with William Brown's help, old grudges apparently dying hard).[193] The federal entomologists claimed that there were hybrids of the red and black ants, proving that the two forms interbred and so could not be distinct species. But most myrmecologists accepted Buren's contention. Two decades later, one of his students showed that genetic evidence conclusively proved that the red and black ants were separate species.[194]

The taxonomic change not only undermined Wilson's theory; it also had a corrosive effect on some of the USDA's claims. In part, the PPC had based its contention that the imported insect might reach the far northern states on the ant's ability to survive cold temperatures in southern South America. But, it was the black ant that pushed into the coldest regions; the red ant stayed in more temperate climes. Since the red and black ants now belonged to separate species, the USDA could no longer use the black ant's tolerance for cold temperatures as evidence that the red ant threatened the entire country.[195] Nor was the insect's irruption across the South fueled by its mutant genes. Blum's fieldwork showed that the ant spread primarily through shipments of nursery stock. "Thus, the story of a species which is running wild must be relagated [sic] to fairy tale status for all time," Blum said.[196] He

190. On Wilson's admission, see M. S. Blum to W. S. Creighton, 8 June 1971; W. F. Buren to W. S. Creighton, 2 July 1971, both in Creighton papers; E. O. Wilson, *The Insect Societies* (Cambridge: Harvard University Press, 1971), 454; E. O. Wilson, *Naturalist* (Washington, D.C.: Island Press, 1994), 117.

191. For Blum's discoveries, see M. S. Blum to W. S. Creighton, 18 November 1970, Creighton papers; M. S. Blum, "Arthropod Defensive Secretions," in *Chemical Releasers in Insects*, ed. A. S. Tahori, *Proceedings of the 2d International IUPAC Congress of Pesticide Chemistry*, vol. 3 (New York: Gordon and Breach Science Publishers, 1971), 163–76.

192. W. S. Creighton to W. F. Buren, 16 July 1971, Creighton papers.

193. For a discussion of the USDA's objections and Brown's role, see "Public Hearing to Determine Whether or Not the Registration of Mirex Should be Cancelled or Amended," F.I.F.R.A. Docket no. 293, transcript pages, 10,813–20, 7 January 1975, box 66, Markin, 1/7/75 file, EDF papers.

194. K. E. Ross and J. C. Trager, "Systematics and Population Genetics of Fire Ants (*S. saevissima* complex) from Argentina," *Evolution* 44 (1990): 2113–34.

195. "Statement of William Franklin Buren," 29 January 1974, Buren file; "Public Hearing to Determine Whether or Not the Registration of Mirex Should be Cancelled or Amended," F.I.F.R.A. Docket no. 293, 14 March 1974, Buren 3/14/74 file, both in box 62, EDF papers.

196. M. S. Blum to W. S. Creighton, 26 January 1969, Creighton papers.

Figure 12. "Fight Fire Ants" bumper sticker. At the outset of the second eradication attempt, the USDA distributed these bumper stickers to steel public resolve against the insect. Creighton thought that they could better be used to encourage the ant by stringing two together into the cheer "Fight, Fire Ants, Fight." (Bumper sticker courtesy of Murray S. Blum.)

wished "everyone would accept the fact that these fire ants are part of the American scene and get on with the business of remedying some real domestic problems."[197]

To the myrmecologists, the ant appeared much more benign than the USDA claimed. They knew that it could be a nuisance—"bad tempered bastards who will bite and sting at the drop of a hat," Creighton said—but suspected that the USDA (aided by Wilson) had exaggerated the danger to increase its funding (and Wilson's professional status).[198] Creighton almost felt sorry for the insect. When the USDA distributed bumper stickers urging the South to "Fight Fire Ants," he suggested that two be strung together: "Fight, Fire Ants, Fight, for after the way that the USDA has treated them, the poor things probably need the encouragement" (fig. 12). But the ant needed little cheering. It had traveled to North America thirty, almost forty years before, spread from Mobile to occupy well over 20 million acres, and withstood the USDA's onslaught for a decade. The ant was a survivor, as Creighton also acknowledged: "When it comes to a showdown," he said, "you put your money on *Solenopsis* and you won't lose."[199]

In Buren's 1972 paper describing the results of his research and the revised taxonomy of the fire ants, by convention the black insect was given the moniker *Solenopsis richteri*. But for the red imported fire ant, Buren chose a name that expressed admiration for the insect.[200] He called the red ant *Solenopsis invicta,* the unconquered fire ant, invoking William Ernest Henley's poem "Invictus":

> Out of the night that covers me,
>   Black as the Pit from pole to pole,
> I thank whatever gods may be
>   For my unconquerable soul.
> In the fell clutch of circumstance
>   I have not winced nor cried aloud,
> Under the bludgeonings of chance
>   My head is bloody but unbowed.

197. M. S. Blum to W. S. Creighton, 1 April 1971, Creighton papers.

198. W. S. Creighton to W. F. Buren, 16 July 1971, Creighton papers (quotation). For Creighton's dim view of the USDA, see W. S. Creighton to R. E. Gregg, 9 March 1968; W. S. Creighton to M. R. Smith, 16 February 1969 and 24 February 1969; W. S. Creighton to W. F. Buren, 17 May 1972; for Blum's views, see M. S. Blum to W. S. Creighton, 1 April 1971; M. S. Blum to W. S. Creighton, 21 September 1971; for Buren's, W. F. Buren to W. S. Creighton, 29 January 1969, all in Creighton papers.

199. W. S. Creighton to W. F. Buren, 16 July 1971, Creighton papers.

200. W. F. Buren, "Revisionary Studies on the Taxonomy of the Imported Fire Ants," *Journal of the Georgia Entomological Society* 7 (1972): 1–26.

Beyond this place of wrath and tears
    Looms but the horror of the shade,
And yet the menace of the years
    Finds, and shall find me, unafraid.
It matters not how strait the gate,
    How charged with punishments the scroll,
I am the master of my fate:
    I am the captain of my soul.[201]

"Let me congratulate you on a brilliant choice of a name for the 'red form,'" Creighton wrote to Buren. "I very much doubt that your head is going to be bloody and I know it is unbowed."[202]

Conclusion

"In an atmosphere of warm emotion," wrote the Alabama journalist George Prentice in 1958, "the value of the tiny, tempestuous fire ant rose and fell with irregular uncertainty."[203] Prentice was commenting on a heated meeting between proponents of the USDA's eradication program and its detractors, but the statement encapsulates the dynamism of the debate over the meaning of the ant generally. The PPC's entomologists wanted to eradicate it; E. O. Wilson insisted that it was a pest, but lambasted the eradication program; J. Lloyd Abbot embraced the eradication ideal, but thought the imported fire ant was an unworthy target; many of the South's university entomologists questioned both the ant's status and the need for eradicating it; conservationists and environmentalists worried that the insecticides did more harm than good—although they offered different reasons for their assessments. One species of ant provoked a panoply of views.

In chapter 2, I traced the construction of the ant's image as a pest. In this chapter, I traced how a different, and contrary, portrait emerged. The players in this latter development did not always agree about the severity of problems that the insect caused, but their opinions clustered around the belief that the ant was at most a nuisance—and the insecticides used to kill it were either a threat to all life that could not be abided or, at least, were used with little regard for their repercussions.

---

201. W. E. Henley, *Poems, 1898* (New York: Woodstock Books, 1993), 119. There remains some controversy over the name. See J. C. Trager, "A Revision of the Fire Ants *Solenopsis geminata* Group (Hymenoptera: Formicidae: Myrmicinae)," *Journal of the New York Entomological Society* 99 (1991): 174; S. W. Taber, *Fire Ants* (College Station: Texas A&M Press, 2000), 12, 25–26.

202. W. S. Creighton to W. F. Buren, 16 July 1971, Creighton papers.

203. G. Prentice, "Fire Ant Program Ordered Stepped up after Hearing," *Montgomery Advertiser,* 2 July 1958, 1.

These ideas, I argue, emerged out of the interactions between a changing natural world and people who spoke from different niches within American society, and who therefore saw the natural world from varying perspectives. The ant, I suggest, changed its behavior in some areas where it had long been established, its density declining along with its tendency to attack crops and vertebrates. At the same time, insecticides followed their own natural laws, killing ants and quail, white-fringed beetles and songbirds. They even slipped into cows' milk.[204] These events and changes were observed by people from different perspectives who therefore interpreted these changes differently. For Buren, Blum, and Creighton, the ant was a spur to attack Wilson: their views reflected their position within a myrmecological community that Wilson was coming to dominate. The vendetta driving them does not invalidate their science, but explains their choice of problems and the energy with which they studied the insect. Science is a human activity, an analysis of the world shaped by our place within it, the resources available, and the viewpoints we carry in our heads. As Creighton told Buren, "No scientist, however devoted, can avoid the personal equation. Not even a computer can arrive at a cold, impersonal scientific conclusion for they are programmed by fallible humans. All that the scientist can do is to hope that his opinions will be like Ivory Soap, 99% pure. The sad fact is that most of us don't make it by about fifty percent."[205]

The South's university entomologists also approached the ant and the insecticides with an axe to grind. They disliked the eradication program and thought that the USDA needed to conduct more research on the ant's biology before sending planes in the air loaded with heptachlor and dieldrin. Studying the ant in places where its population was declining and using methods that highlighted the insect's positive attributes, the southern entomologists found an insect that little resembled the one described in the USDA's press releases. Results from these studies then became weapons in a battle about the proper aims and practices of entomology. Again, the point is not that the personal and political considerations invalidated the studies. Professional worries gave the research an urgency and the methods highlighted the ant's benign characteristics; but the entomologists really did see declines in the ant's population, really did see it eat other insects, and really learned that many of the claims about the ant's depredatory ways

204. The USDA acknowledged that the insecticides used to eradicate the ant contaminated milk in R. N. Dopson to C. C. Fancher, 12 June 1961, fire ant file, box 93, PPC papers.

205. W. S. Creighton to W. F. Buren, 26 November 1970, Creighton papers.

were exaggerations. Their research was—and remains—a useful corrective to the PPC's hyperbole, although Newsom might have substituted his own exaggerations for the USDA's.

Conservationists and environmentalists built on the work of the university entomologists, but pushed it in different ways. For conservationists, the profligate use of insecticides proved that the gospel of efficiency was out of kilter and suggested that U.S. public policy was skewed. Environmentalists, too, worried that public policy did not reflect the public good, but not because the USDA perverted the conservation ideal. Robert Rudd, Harold Peters, and Rachel Carson saw the fire ant eradication program as the epitome of the "production fetish" and asked Americans to look at nature differently: not as resources to be exploited, but as a nexus of relationships on which all our lives depend. The fire ant could safely enter this nexus, but the insecticides never could—one was natural, the other unnatural. As much as the USDA sought to refine the ambiguous image of the fire ant to make the insect into a fiend that needed to be eradicated, Carson stripped the insect of ambiguity to show that it was a part of the natural world while insecticides represented human hubris and autocratic governance. Both the conservationists' and environmentalists' opinions of the insect, the insecticides, and the eradication program emerged out of the interaction between the natural world and the ideas of those who represented different interests in American society.

Chapter Four: The Fire Ant Wars, 1958–1983

Between the idea
  And the reality
Between the motion
  And the act
Falls the Shadow

T. S. Eliot, "The Hollow Men"

By the late 1950s, a number of ideas about the red imported fire ant and the insecticides used to kill it had taken shape. The various responses to the ant were more than just ideas, however. They were blueprints for action. If *Solenopsis invicta* was among the worst ant pests in the Western Hemisphere and the insecticides killed some wildlife, but not much, then the eradication program needed to go forward full force. If, on the contrary, the ant was as helpful as it was harmful, and the insecticides decimated wildlife and poisoned humans, then no eradication program was warranted; either smaller control programs could be undertaken, or the ant could be left alone, free to blend into the southern ecosystem.

This chapter moves out of the laboratory and away from the farm—although not too far from either—to see how ideas about the red imported fire ant and the insecticides were translated into public policy.[1] When the U.S. Department of Agriculture (USDA) first announced the eradication campaign in 1957, it seemed that critics' objections might be taken into consideration during the policymaking process and a program acceptable to most people put into place. By the middle of 1958, though, that hope had disappeared, replaced by acrimony: the fire ant wars. Two sides coalesced and they battled in the press, the halls of Congress, and state legislatures.

About seven years after the eradication program began, it looked dead, and officials in the USDA moved to cease funding for the program. But a concatenation of events resuscitated it. A new, seemingly safe insecticide, a new under secretary of agriculture, and renewed interest by the states combined to spark a second eradication campaign. This program, like the one before it, met stiff opposition from

---

1. Compare B. Latour, *Science in Action: How to Follow Scientists and Engineers through Society* (Cambridge: Harvard University Press, 1987).

those unconvinced that eradication of the ant was called for or could be done safely. Now, however, there were new forums for making decisions about how America should deal with this invading insect. In 1970, President Richard Nixon created the Environmental Protection Agency (EPA), and when questions arose about the eradication program, it was this agency that was charged with deciding the worthiness of the campaign. After a protracted battle, those opposed to the new eradication effort triumphed. But even as the eradication program ended, the ants were changing in unexpected ways, mocking the hope that they would become well-mannered members of the southern ecosystem.

Prologue to War

Although the red imported fire ant and the insecticides used to kill it inspired heated debates, the battle between the USDA and those who became its opponents was not inevitable. For a time, compromise seemed possible. While conservation groups voiced their concerns to the press, they were nonetheless willing to cooperate with the agency. Waldo Lee Popham, an eradication proponent and deputy administrator of the Agricultural Research Service (ARS), convened a meeting with wildlife advocates to discuss their fears, the USDA established a liaison with the U.S. Fish and Wildlife Service (USFWS), and, in early 1958, federal entomologists promised to protect birds and other animals.[2] This spirit of cooperation was convincing to wildlife biologist Harold Peters, who encouraged other wildlife biologists to withhold criticism.[3] Peters was so supportive that the USDA considered hiring him as a consultant and, even after he joined the National Audubon Society, Lamar J. Padget, C. C. Fancher's lieutenant in charge of the fire ant program, thought that his reports would be "fair and unbiased."[4] Anxiety over the eradication program was not eliminated: Walter Rosene of the USFWS refused to endorse the eradication program until he was certain

2. On Popham's meeting, see W. L. Popham to S. H. Ordway Jr., 26 December 1957, regulatory crops-fire ant file, box 752, General Correspondence, 1 (UD), Agricultural Research Service papers, RG 310, National Archives and Records Administration, College Park, Md. (hereafter ARS papers). On the liaison between the two bureaucracies, see M. R. Clarkson to D. H. Janzen, 20 December 1957, fire ant files (1957), Plant Pest Control Division papers, RG 463, National Archives and Records Administration, College Park, Md. (hereafter PPC papers).

3. W. F. Barthel to C. H. Gaddis, 12 March 1958, fire ant file, box 30, Entomology Research Division: General Correspondence, 1954–1958, 5 (UD), ARS papers.

4. On the PPC contemplating hiring Peters, see L. J. Padget to E. D. Burgess, 27 November 1957, fire ant files (1957); the quotation is from L. J. Padget to C. C. Fancher, 3 March 58, imported fire ant file, box 198, both in PPC papers.

that it did not kill wildlife, and other officials within the service fretted about the potential damage to the South's game animals.[5] But criticism of the program seemed restricted to the interdepartmental relationship between the USDA and the USFWS. In early 1958, when federal wildlife biologists found a number of dead animals in southwest Georgia after the USDA sprayed heptachlor and dieldrin, they did not make it public nor did they criticize the agricultural program.[6] There was, one USFWS official noted, a "gentleman's agreement" to keep discussions private.[7]

The situation, though, was not stable, and whatever goodwill existed at the end of 1957 began to dissolve in the spring of 1958 when the Alabama Department of Conservation found the dead animals in Autauga County following a spraying of heptachlor (see chapter 3, fig. 11). The USFWS counseled patience, but other conservationists called for an immediate end of the program.[8] The state Department of Conservation and Peters took the matter to the press, while Auburn wildlife biologists published reports of wildlife deaths over the objection of USDA officials, who felt that their data were scanty.[9] Entomologists with the USDA's Plant Pest Control Division (PPC) felt betrayed, and I think that this feeling accounts for much of the acrimony in the ensuing fire ant wars. Supposed allies suddenly turned against them, and the USDA entomologists reacted sharply.[10] They no longer told wildlife groups where they would next be spraying for the

5. On Rosene, see C. E. Carlson to D. H. Janzen, 6 October 1958, fire ant file, box 22, Wildlife Research Correspondence, 254 (UD), U.S. Fish and Wildlife Service papers, RG 22, National Archives and Records Administration, College Park, Md. (hereafter USFWS papers). On the USFWS's worries, see E. D. Burgess to C. C. Fancher, 16 December 1957, fire ant files (1957), PPC papers; Minutes of the 37th Interdepartmental Meeting, 13 March 1958, interdepartmental pest control file, box 77, Entomology Research Division: General Correspondence, 1954–1958, 5 (UD), ARS papers.

6. On the findings and the decision to keep them out of the press, see C. E. Carlson to D. L. Leedy, 18 March 1958; D. H. Janzen to Regional Directors, 18 March 1958; D. L. Leedy to A. Nelson, 2 May 1958, all in fire ant file, box 22, Wildlife Research Correspondence, 254 (UD), USFWS papers.

7. D. L. Leedy to W. W. Dykstra, 22 July 1958, fire ant file, box 22, Wildlife Research Correspondence, 254 (UD), USFWS papers.

8. For the service's response, see D. H. Janzen, "Effects of the Fire Ant Eradication Program upon Wildlife," 25 May 1958, fire ant file, box 30, Entomology Research Division: General Correspondence, 1954–1958, 5 (UD), ARS papers.

9. On disputes between Auburn and the PPC, see W. O. Owen to C. C. Fancher, 15 December 1958, fire ant file, box 1981; L. J. Padget to E. D. Burgess, 30 March 1959, fire ant file, box 213, both in PPC papers.

10. For the belief that the USDA's opponents were unethical in violating the "gentleman's agreement," see L. F. Curl to K. T. Karabatsos, 4 June 1959, fire ant file, box 213, PPC papers.

ant.[11] Rosene became, in the words of a USFWS official, "persona non grata."[12] As Rosene, Dan Lay, and other wildlife biologists found more dead animals and calls for ending the program became increasingly strident, the USDA dug in its heels even deeper. Popham insisted, "It seems to me that people who have ants should be privileged to get rid of them if they want to. We feel that the best answer to some of the criticism that has been made of the program is to keep moving ahead, utilizing the best procedures that we have available and taking full advantage of the advice and counsel of local groups who are interested in the work and have property to protect."[13]

The diversity of opinions inspired by the ant, the insecticides, and the eradication ideal resolved into two mutually exclusive positions: any middle ground disappeared. Padget came to view Peters as a "rabble rouser" and wildlife biologists whispered vile rumors about PPC entomologists: after *Life* magazine published a picture of a woman's leg covered in welts from fire ant stings, the USDA's opponents speculated that the woman was a prostitute whom PPC entomologists had gotten drunk and then laid across a fire ant mound to fake the evidence that the fire ant was a pest that they otherwise lacked.[14] Meanwhile, some eradication advocates may have resorted to bullying and intimidation. Lay and Rosene, for example, both reported that people associated with the program tried to have them fired and, in Rosene's case, evidence exists to support the charge. Wildlife officials received a letter from William Blasingame, the Georgia state entomologist and one of the PPC's closest allies, asking that Rosene be removed from the program.[15]

11. W. O. Owen to E. D. Burgess, 28 January 1958, fire ant file, box 197, PPC papers; W. W. Dykstra to D. L. Leedy, 3 August 1958, fire ant file, box 22, Wildlife Research Correspondence, 254 (UD), USFWS papers; W. E. Westlake to S. A. Hall, 18 December 1959, pesticides chemical research laboratory file, box 5, Entomology Research Division: Entomology Director's Correspondence, 1959–1965, 1055 (A1), ARS papers.

12. C. E. Carlson to D. H. Janzen, 6 October 1958, fire ant file, box 22, Wildlife Research Correspondence, 254 (UD), USFWS papers.

13. W. L. Popham to W. G. Cowperthwaite, 28 October 1958, regulatory crops, 1958, fire ants file, box 943, Entomology Research Division: General Correspondence, 1954–1966, 1 (UD), ARS papers.

14. Padget's quote is from L. J. Padget to E. D. Burgess, 30 October 1958, fire ant file, box 197, PPC papers. The story about the prostitute is from an interview with John L. George, 16 April 1998, notes in author's possession. For the magazine article, see "Fire Ant Plague," *Life* 45 (14 July 1958): 117–18.

15. On Lay's troubles, see W. Rosene to C. Cottam, 16 December 1958, folder 773; C. Cottam to R. Carson, 11 April 1962, folder 774, both in Rachel Carson papers, YCAL 46, Beinicke Rare Book Library, Yale University (hereafter Carson papers); for Blasingame's attempt, see W. E. Blasingame to H. E. Talmadge, 23 May 1958; A. J. Suomela to H. E. Talmadge, 6 June 1958, both in fire ant file, box 22, Wildlife Research Correspondence, 254 (UD), USFWS papers.

Rosene also claimed that USDA entomologists followed him wherever he drove, tapped his phone, and spread heptachlor pellets on the path from his hotel room in Georgia to his car, although there is no documented proof of this harassment.[16]

In circumstances such as these, there was no room for compromise, as wildlife biologist and entomologist Harlow B. Mills discovered in late 1958.[17] Tapped by the USDA to bridge the growing divide between the department and its opponents, Mills found fault with both sides. He said that the conservation groups had not been "very moral" to release notices of wildlife deaths without alerting the USDA first and blamed them for spreading "misleading and sensational propaganda."[18] Mills also castigated PPC entomologists for their "domineering, arrogant" attitude and misplaced faith in the eradication ideal, which he called "unrealistic."[19] Mills enjoined both sides to forget past hurts and work together, but his pleas went unheeded; PPC entomologists and their opponents continued to look on each other warily. Padget told Mills, "It is our opinion that the work will continue along this line at least for the foreseeable future for nothing is presently in sight that would appreciably alter the current operations."[20] The USDA's opponents also refused to budge. Clarence Cottam, head of the Welder Wildlife Foundation, for example, told Mills, "I, personally, should like very much to take a more moderate course on this controversial issue and, if we could get any support from Agriculture, I believe I would be one of the first to approach it from this angle. At this time, however, it seems to me we have no alternative than to meet the control people on their own grounds. . . . In fighting for great issues, as on the field of battle, one cannot sit calmly and scientifically consider each minute detail and expect the opposition to follow this same course. American democracy was not won on such a basis and I don't believe it can be retained with this approach."[21] With the possibility of compromise gone, the groups spent the next two decades in battle, each attempting to see its preferred policy made real.

16. Interview with Walter Rosene, Gadsden, Alabama, 24 August 1999, audiotapes and notes in the possession of the author.

17. For Mills's biographical data, see "Harlow B. Mills, 1906–1971," folder 20, box 9, Systematic Entomology Laboratory papers, RG 7323, Smithsonian Institution Archives, Washington, D.C.

18. H. B. Mills to W. L. Popham, 29 December 1958, man and animals imported fire ant file, box 1, Entomology Research Division: Entomology Director's Correspondence, 1959–1965, 1055 (A1), ARS papers.

19. Ibid.

20. L. J. Padget to H. B. Mills, 25 February 1959, fire ant file, box 213, PPC papers.

21. C. Cottam to H. B. Mills, 9 October 1959, Walter Rosene papers, Auburn University.

The Fire Ant Wars

By mid-1958, the PPC entomologists found themselves fighting a war on two fronts, battling both the red imported fire ant and those opposed to the program. The battle against the ant involved converted World War II bombers dropping dieldrin and heptachlor over southern fields and farms and suburbs; the battle against the South's university entomologists, wildlife biologists, and the nascent environmental movement took place in the press, in the halls of Congress, and in state legislatures. During the first few years of the eradication campaign, the USDA seemed to be winning on both fronts. The ant had spread widely, but the insecticides were powerful and in some places, PPC entomologists reported, 98 percent of the insects died after the fields were sprayed the first time.[22] Meanwhile, the USDA battled its critics from a position of power. The PPC entomologists had initiated the program before any substantial opposition could be organized, and the iron triangle between the federal scientists, congressional allies, and lobbyists such as the Farm Bureau protected the program.

One of the tools eradication opponents used to try to break this iron triangle was the press. The Audubon Society's Harold Peters, especially, worked to swing public opinion against the fire ant eradication program. We need to "combat the flagrant propaganda of USDA," he told Daniel Leedy, chief of the USFWS's wildlife research branch.[23] Peters worked with journalists who wrote articles for *True, Reader's Digest,* and *Parents* magazine critical of the fire ant program or the profligate use of insecticides.[24] He gave talks to gardening groups, conservation organizations, and local reporters.[25] And he appeared on a television show with Ed Komarek, protégé of Georgia quail biologist Herbert Stoddard, arguing that the imported fire ant had little or no effect on quail.[26] Others also spread opposition to the eradication ideal through the press. The Alabama Department of Conservation enlisted a publicist to help in its attempt to stop the program, and Cottam and Mobile nurseryman J. Lloyd Abbot, too, made their opinion public. Only the

22. On the efficiency of the eradication program, see "Appraisal Survey of the Cooperative Fire Ant Program," 12 October 1960, fire ant files (1960), PPC papers.

23. H. S. Peters to D. L. Leedy, 22 July 1958, fire ant file, box 22, Wildlife Research Correspondence, 254 (UD), USFWS papers.

24. For Peters's work with journalists, see "Report of Field Investigations in Tennessee, Mississippi, Louisiana, Texas, and Alabama," 1–21 May 1959; H. S. Peters to C. W. Bucchesiter, 31 December 1959; H. S. Peters to C. Callison, 14 May 1960, all in Peters file, National Audubon Society papers, New York Public Library (hereafter NAS papers).

25. See Peters's accounts of his work in Peters file, NAS papers.

26. H. S. Peters to J. H. Baker, 19 March 1959, Peters file, NAS papers.

USFWS and Rachel Carson were reluctant to criticize the USDA publicly—until Carson published *Silent Spring*.[27]

In late 1958, federal entomologists prepared to spray Monroeville, Alabama, careful not to give notice to wildlife biologists. Officials with the Alabama Department of Conservation caught wind of the plans, though, and sent game wardens to quietly investigate.[28] The wardens collected a number of dead animals, then alerted journalist Jerry Hornsby of the *Alabama Journal*, who reported that a survey of thirty-six homeowners found 697 dead chickens, 10 roosters, 20 turkeys, 11 cats, 7 puppies, and 2 ducks.[29] On another occasion, a National Wildlife Federation member posed as a teacher to obtain an advance copy of *The Fire Ant on Trial*, a filmstrip produced by the USDA in 1958 to counteract bad publicity. The federation member airmailed the film to the Alabama Wildlife Federation and the Alabama Department of Conservation so that the USDA's critics could preview the film with local journalists. The hope was that the conservation groups could point out inaccuracies and have the press debunk the movie before it received wide release.[30] Other conservation groups saw the film after its release and excoriated it in the *New York Times*.[31]

The public relations savvy of those opposed to the eradication program, though, was no match for the USDA. Over the century since its founding, the department had grown into one of the most powerful government agencies; it was, according to political scientists Theda Skocpol and Kenneth Finegold, an island of strength in an ocean of administrative weakness during the first half of the twentieth century, and even as the federal government's capacities expanded during the Great Depression and the Cold War, the USDA retained a great deal of power.[32] (In 1940, the USDA received 40 percent of all federal funds earmarked for research; by 1962, that was down to 1.7 percent, or about 5 percent of non-Defense research monies. But its funding still grew,

27. On Carson's reluctance, see L. Lear, *Rachel Carson: Witness for Nature* (New York: Norton, 1997), 347.

28. L. F. Curl to K. T. Karabatsos, 4 June 1959, fire ant file, box 213, PPC papers.

29. J. Hornsby, "Fire Ant Poison Plays Havoc in City Limits of Monroeville," 30 October 1958, fire ant file, SG 8837, Alabama Department of Agriculture and Industry papers, Alabama Department of Archives and History, Montgomery, Ala. (hereafter ADAI papers).

30. H. S. Peters to J. H. Baker, 19 March 1959, Peters file, NAS papers.

31. "U.S. Gets Protest on Fire Ant Film," 26 April 1959, folder 584, Carson papers. See also I. R. Barnes, "Open Letter to Ezra Taft Benson," *Natural Food and Farming* 6 (1959): 17.

32. T. Skocpol and K. Finegold, "State Capacity and Economic Intervention in the Early New Deal," *Political Science Quarterly* 97 (1982): 271. On the development of the USDA, see A. H. Dupree, *Science in the Federal Government: A History of Policies and Activities* (Baltimore: Johns Hopkins University Press, [1957] 1986), 149–83; G. L. Baker, *Century of Service: The First 100 Years*

from $80 million to $12.5 billion.)[33] With the extension service and connections to the American Farm Bureau, the department's reach extended into almost every county in America.[34] Using this extensive network of contacts and skills honed over decades, the USDA promoted its vision of the ant and the need for the eradication campaign. Assigning a publicist to promote the program was one method; producing *The Fire Ant on Trial* was another. And as the program developed, the USDA redoubled its publicity efforts. In December 1958, W. O. Owen, head of the PPC's operations in Alabama, told Fancher that he was "presently working local information campaigns on the fire ant to keep the local people advised" and recommended hiring "an outstanding writer of much capability to throw off the activities that are now taking place in opposition to the fire ant program."[35] Throughout the late 1950s and into the 1960s, the agency sent press releases to local papers detailing the need for eradication of the ant and the safety of the insecticides.[36] The PPC also benefited from its alliance with the chemical industry. In May 1958, the National Agricultural Chemical Association recorded Herbert Stoddard saying, among other things, "We think with the imported fire ant that up to thirty to fifty percent of the [quail] nests might be destroyed." The industry group put the interview on phonographs that it sent to radio stations throughout the South.[37] The USDA's publicity machine seemed successful. The Alabama Farm Bureau found that 802 of the 803 farmers who responded to a survey supported the eradication campaign. (The total number asked was 1,020.)[38] Fancher said, "Farmers and urban dwellers alike, whose property has been treated for the imported fire ant, have expressed more appreciation than they have for any other program in which this division has had a part."[39]

---

of the United States Department of Agriculture (Washington, D.C.: USDA, 1963); D. E. Hamilton, "Building the Associative State: The Department of Agriculture and American State-Building," *Agricultural History* 64 (1990): 207–18; G. Hooks, "From an Autonomous to a Captured State Agency: The Decline of the New Deal in Agriculture," *American Sociological Review* 55 (1990): 29–43.

33. U.S. Congress, Subcommittee on Agriculture, Committee on Appropriations, *Department of Agriculture Appropriations: FY 1963*, 87-2, 1962, 371.

34. On the relations with the extension service and the Farm Bureau, see S. R. Berger, *Dollar Harvest: The Story of the Farm Bureau* (Lexington, Mass.: Heath Lexington Books, 1971), 89–126.

35. W. O. Owen to C. C. Fancher, 15 December 1958, fire ant file, box 198, PPC papers.

36. See, for example, V. E. Weyl to C. C. Fancher, 17 December 1959, fire ant file, box 213; V. E. Weyl to E. D. Burgess, 11 April 1960, fire ant files (1960), both in PPC papers.

37. H. S. Peters to R. Clement, 14 December 1959, Peters file, NAS papers.

38. D. R. Shepherd to G. H. Bradley, 30 January 1962, information-12-talks, speeches file, box 134, PPC papers.

39. C. C. Fancher to J. F. Spears, 7 June 1960, fire ant files (1960), PPC papers.

Popular approval for the program—broad, if not universal—translated into congressional approbation. Jamie Whitten, representative from Mississippi, was reported to have said, "It is the responsibility of the Congress to see that the farmers are supported to their benefit and until they get adverse reports from the farmers on the program they do not have to worry about fire ant money."[40] Still, the PPC officials could not rely on the program's popularity to assure continued funding, for their opponents were cultivating their own political allies and working to cut off appropriations. In the middle of 1958, for example, some of Alabama's congressional delegation suggested a moratorium on the program after hearing of the ant research by Auburn University's Hays brothers and the wildlife killed by insecticides.[41] To head off these attacks, agricultural officials dissembled. Testifying before Whitten's congressional committee, USDA officials explained that the wildlife deaths were atypical. An accidental spill on a landing strip was claimed to be responsible for killing the animals in Autauga County, while rain was said to be responsible for the dead animals that Dan Lay had found in Hardin County, Texas. Three inches of rain followed the insecticide treatments, the officials testified, and the chemicals seeped into puddles where they accumulated, and then poisoned the animals that drank the water.[42] But, as Clarence Cottam said about other USDA dismissals of wildlife deaths, the statements "were prepared either in alarming ignorance of the facts or with a lack of intellectual integrity."[43] The landing strip was three miles from where the dead animals had been found and the rain came nineteen days after the application of pesticides in Texas.[44] Eighty-seven percent of the dead animals had been collected by then.[45] Although the USFWS's Daniel Janzen complained about such factual inaccuracies to the USDA's top brass, the department insisted on its interpretation of the events.[46] And Congress continued its funding of the program.

Rebuffed by Congress, wildlife biologists and environmentalists pinned their hopes for reform on the USFWS. The agency had suffered a difficult decade after the end of World War II, but in the mid-1950s

---

40. J. S. Riss, memo, 27 March 1959, fire ant file, box 213, PPC papers.

41. On opposition to the program on the part of Alabama, see "Solons Ask Evaluation of Ant Plan," *Mobile Press-Register*, 23 September 1958, 1.

42. D. H. Janzen to B. T. Shaw, 17 July 1959, pesticides, 1949–1960 file, box 24, Wildlife Research Correspondence, 254 (UD), USFWS papers.

43. C. Cottam to M. Shurtleff, 14 October 1958, fire ant file, box 197, PPC papers.

44. D. H. Janzen to B. T. Shaw, 17 July 1959, pesticides, 1949–1960 file, box 24, Wildlife Research Correspondence, 254 (UD), USFWS papers.

45. Ibid.

46. Ibid.; B. T. Shaw to E. T. Benson, 25 July 1960, fire ant files (1960), PPC papers.

underwent a renaissance, with new leadership, more power, and a new agenda—"Operation Life," a program to elucidate the mysteries of "life sufficiently to manage it in accordance with the best interests of our expanding population."[47] Wildlife research, long neglected in favor of programs directed at killing predators, grew.[48] Officials expressed interest in studying the relationship between wildlife and increasingly common insecticides, eventually making the fire ant program the "major field project on pesticide-wildlife relations."[49] The community of conservation activists hoped that the USFWS would lead the charge against the USDA. Cottam told Janzen, "The wildlife interests of the country will be looking to the service to take objective leadership."[50] The Audubon Society lobbied Congress to increase the USFWS's budget for research on insecticides to $25 million dollars.[51] And in 1958 many rallied around Iowa representative Leonard Wolf's bill requiring the USDA to consult with the USFWS before undertaking any eradication program. It seemed a good hammer for breaking the iron triangle that protected the USDA's eradication campaigns.[52]

Relying on the USFWS, though, proved dangerous to conservationists' interests and, in the end, disappointing. A congressional committee sympathetic to the wildlife groups conducted hearings on the Wolf bill, and progress at first seemed positive. Representative John Dingell of Michigan severely criticized the USDA and closely questioned Popham about a report that the USDA had tried to have ornithologist George Wallace disciplined for objecting to an eradication program in the Midwest.[53] Watching the proceedings, James Dewitt reported that

47. On the USFWS's changing fortunes, see D. L. Allen to R. L. Leffler, 14 March 1957, Allen file, box 1, Correspondence of Assistant Secretary for Fish and Wildlife Ross L. Leffler, 790 (A1), Secretary of Interior papers, RG 48, National Archives and Records Administration II, College Park, Md. (hereafter USDI papers); U.S. Congress, Subcommittee on Fisheries and Wildlife Conservation, Committee on Merchant Marine and Fisheries, *Reorganization Hearings*, 84-2, 1957; J. N. Clarke and D. C. McCool, *Staking Out the Terrain: Power and Performance among Natural Resource Agencies* (Albany: State University of New York Press, 1996), 107–25. For Operation Life, see "Ten-Year Program for Wildlife Research," 1957, file 23, box 8, Bird and Mammal Laboratory Records, 1885–1971 (USFWS) papers, RG 7171, Smithsonian Institution Archives, Washington, D.C.

48. "Responsibilities and Objectives of the Branch of Wildlife Research," n.d., historical file, box 6, Office Files of Dr. Frederick C. Lincoln, 1917–1960, 253 (UD), USFWS papers.

49. A. V. Tunison to Regional Director, Atlanta, 19 December 1957, fire ant file, box 22, Wildlife Research Correspondence, 254 (UD), USFWS papers (quotation).

50. C. Cottam to D. H. Janzen, 13 October 1958, fire ant file, box 22, Wildlife Research Correspondence, 254 (UD), USFWS papers.

51. U.S. Senate, Subcommittee on Interstate and Foreign Commerce, Committee on Commerce, *Hearings*, 85-2, 1958, 17.

52. C. Callison to C. Cottam, January 20, 1960, Cottam file, NAS papers.

53. U.S. Congress, Subcommittee on Fisheries and Wildlife Conservation, Committee on Merchant Marine and Fisheries, *Coordination of Pesticide Programs*, 86-2, 1960, 65–77.

"Popham's bald head turned as red as a baboon's fanny, and he said—as piously as a parson p——g in a snowdrift—'Oh, we'd *never* try to get anybody fired.' Wonder if Dan Lay, Les Glasgow, and some others will believe this statement."[54] But, there was no joy among the USDA's critics after Janzen testified. Janzen told the committee that the bill was "premature" and "impracticable." "With few exceptions we are not yet in a position to recommend control methods that do not have deleterious effects on these resources," he said.[55] The director of the Florida Fish and Game Commission thought that Janzen's "testimony may kill the bill."[56] He was wrong, but the testimony certainly did not help. The bill passed out of committee but died later, probably killed by agricultural lobbying.[57]

Janzen's testimony was not out of character. Since nominating the USFWS to lead the conservationsists' efforts, Cottam, Peters, and others had complained that federal wildlife officials were "spineless" for refusing to publicly confront the USDA.[58] The USFWS, much like the National Audubon Society, was in a delicate position. Adjusting to the social and demographic changes that gave birth to the environmental movement, the service remained shackled by a past allegiance to conservation and a history of weakness. The USFWS—as much as the PPC—was seeking to increase its bureaucratic status, but its leaders were not as skilled as the bureaucrats in the PPC. Thus, in 1959, Janzen called for an immediate end to the fire ant eradication program, but his effort petered out within the year. The service then relapsed into its favored practice of seeking to slowly increase its power by emphasizing the need for more research and refusing to antagonize agriculture officials. The federal wildlife officials with the USFWS were so fearful of controversy that after Blasingame tried to get Rosene fired, they blamed their wildlife biologist for antagonizing the USDA![59] (Disillusioned, Rosene left the service a few years later to write a book

54. J. B. DeWitt to W. Rosene, 4 May 1960, Rosene papers. Elision and emphasis in original.

55. U.S. Congress, Subcommittee on Fisheries and Wildlife Conservation, Committee on Merchant Marine and Fisheries, *Coordination of Pesticide Programs*, 86-2, 1960, 53.

56. H. S. Peters to C. Callison, 14 May 1960, Peters file, NAS papers.

57. C. J. Bosso, *Pesticides and Politics: The Life Cycle of a Public Issue* (Pittsburgh: University of Pittsburgh Press, 1987), 106.

58. C. Cottam to R. Carson, 17 January 1963, file 1455, Carson papers (quotation). For other examples, see R. L. Burnap to L. A. Parker, 20 June 1958; H. S. Peters to D. L. Leedy, 22 July 1958; H. S. Peters to W. Dykstra, 31 July 1958; C. Cottam to A. L. Nelson, 2 October 1958; R. L. Burnap to L. A. Parker, 17 December 1958, all in fire ant file, box 22, Wildlife Research Correspondence, 254 (UD), USFWS papers; see also J. H. Baker to C. Cottam, 14 December 1958, Cottam file, NAS papers.

59. W. W. Dykstra to D. L. Leedy, 3 August 1958, fire ant file, box 22, Wildlife Research Correspondence, 254 (UD), USFWS papers.

on quail biology that updated Stoddard's classic work.)[60] Given the USFWS's cautious approach, opponents of the USDA had little hope of using the service to break the iron triangle protecting the fire ant program.

In his study of opposition to nuclear power, historian Thomas Wellock argues that environmental activists who found it difficult to change federal laws turned to state governments—a "backdoor" to regulation, Wellock calls it.[61] Not all of those opposed to the fire ant eradication program were environmentalists, but they also adopted this strategy, finding state governments more accessible than the federal government, and more easily persuaded. In Mississippi, Ross E. Hutchins blocked the state from contributing funds to the fire ant program, and some wildlife biologists in Georgia similarly tried to stop Georgia from joining the campaign.[62] Officials with the Alabama Department of Conservation also tried to convince their state to withdraw from the program—a development that would "be a severe blow to the entire eradication program inasmuch as Alabama is probably the most highly infested state and had previously supported the program," an ARS audit of the fire ant program admitted.[63]

The effort in Alabama came to a head in 1959, when the state's biennial budget was being written. J. Lloyd Abbot coordinated the opposition to the fire ant program, drafting Frank S. Arant of Auburn, Rosene, DeWitt, and a host of others to testify against the program.[64] Red Bamberg, Alabama's agriculture commissioner, organized a carpool of farmers to support continued funding.[65] The hearing was heated and the battle spilled over into a restaurant after the meeting, where Bamberg told Rosene and DeWitt that he planed to wage "war" against

60. Interview with Walter Rosene, Gadsden, Alabama, 24 August 1999, audiotapes and notes in the possession of the author. The book is W. Rosene, *The Bobwhite Quail, Its Life and Management* (New Brunswick: Rutgers University Press, 1969).

61. T. R. Wellock, *Critical Masses: Opposition to Nuclear Power in California, 1958–1978* (Madison: University of Wisconsin Press, 1998), 114–46.

62. On Hutchins, see L. J. Padget to C. C. Fancher, 19 August 1960, fire ant files (1960), PPC papers; on Georgia, see R. E. Tyner to J. Duncan, 26 October 1959, fire ant file, box 213, PPC papers; P. Daniel, "A Rogue Bureaucracy: The USDA Fire Ant Campaign of the Late 1950s," *Agricultural History* 64 (1990): 111.

63. Internal Audit, "Report on Imported Fire Ant Eradication Program in Four Southern States," 11–13 June 1959, fire ant file, box 61, PPC papers.

64. W. C. Younger to E. V. Smith, 5 October 1959; J. L. Abbot and J. E. McMurphy Jr., to J. W. Smith, 14 October 1959; E. V. Smith to W. C. Younger, 14 October 1959; F. S. Arant to the Ways and Means Committee, Alabama House of Representatives, 7 October 1959; J. L. Abbot to R. B. Draughon, 28 October 1959, all in box 2, Ralph B. Draughon papers, RG 107, Auburn University Archives (hereafter Draughon papers).

65. See the correspondence in fire ant file, SG 8837, ADAI papers.

the Alabama Department of Conservation.[66] But this time, Abbot and the rest of the USDA's opponents won the day. The governor sharply criticized Bamberg and the state's Ways and Means Committee killed funding for the fire ant program.[67] In 1961, the Alabama Wildlife Federation lauded Abbot as its conservationist of the year and Abbot returned the compliment, praising the federation for standing up to "bureaucracies."[68] He then asked Auburn to find out if the white-fringed beetle was really a pest, or another figment of the USDA's imagination.[69]

The celebration, though, was short-lived. Despite some setbacks, the PPC was in firm control of the political machinery that authorized the eradication campaign. The same year that Abbot was fêted, Alabama appropriated funds to rejoin the program. (Abbot was out of the country at the time of the hearings; Fancher testified about the program's positive aspects.)[70] The USDA also overcame opposition in other states. Florida, which had quit funding the program as well, rejoined when Alabama did; the Georgia Farm Bureau helped to coordinate the ouster of the wildlife biologist opposed to the program in the state, as J. Phil Campbell, the Peach State's commissioner of agriculture, happily reported to Popham.[71] The PPC also used creative financing to maintain the program. The law required that states and locals contribute 50 percent of the money to the eradication program, but Alabama's agricultural commissioner conceded, "This program, with most of the funds coming from the federal government, is being operated primarily by the federal government. The state is furnishing a very small portion of the money that is being used and our employees our working under the supervision of the federal men."[72] Fancher arranged it—with approval from his supervisors—so that spraying continued in Mississippi, which contributed only a token amount, and Texas, which contributed

66. E. C. Carlson, memo, 9 October 1959, fire ant file, box 22, Wildlife Research Correspondence, 254 (UD), USFWS papers.

67. E. C. Carlson, memo, 9 October 1959, fire ant file, box 22, Wildlife Research Correspondence, 254 (UD), USFWS papers.

68. J. L. Abbot to R. L. Leffler, 17 January 1961, Abbot file, box 1, Correspondence of Assistant Secretary for Fish and Wildlife Ross L. Leffler, 790 (A1), USDI papers.

69. J. L. Abbot to R. B. Draughon, 11 July 1961, box 2, Draughon papers.

70. On Abbot, see J. L. Abbot to R. Carson, 25 September 1961, folder 1586, Carson papers. On Fancher, see W. A. Ruffin to D. R. Shepherd, 21 July 1961, appropriations (budget) hearing file, box 84, PPC papers.

71. On Florida, see C. C. Fancher to E. D. Burgess, 28 March 1960, fire ant files (1960), PPC papers; on Georgia, see J. P. Campbell to W. L. Popham, n.d. (1959), fire ant file, box 213, PPC papers.

72. A. W. Todd to J. P. Henderson, 18 April 1958, fire ant file, SG 8836, ADAI papers.

no funds.[73] By 1960, the PPC had spent $2,426,237 while states, local governments, and landowners had contributed only $1,855,686.[74] The eradication program rolled on, as PPC entomologists treated 1,974,436 acres between the fall of 1957 and the spring of 1960.[75]

The Eradication Campaign Grinds to a Halt

The USDA, historian Pete Daniel notes, considered attacks on the fire ant program a "public relations" problem.[76] Federal entomologists saw no substance to the criticisms, so, rather than review their policies and procedures, they responded to the criticisms by coordinating a publicity campaign to overcome their opponents' contumacy and maintain support for the federal program. "We have an effective method of dealing with the fire ant," Popham asserted in July 1958.[77] Daniel finds this intransigence hard to comprehend. "Even with thirty years' perspective it is difficult to understand the bureaucratic arrogance of the ARS," he wrote in 1990.[78] The trouble, I think, stems from seeing hubris as the sole force motivating the actions of the federal scientists, whereas a more complex mélange of emotions was at work. First, the PPC entomologists saw their critics as treacherous, and so gave their views little credence. Second, the federal scientists were insecure about the future of their relatively new bureaucracy: fanatic orthodoxy is often rooted in doubt, not confidence. They had staked the division's survival on their ability to eradicate insects, an audacious and still unproven dream. I suspect PPC officials worried that admitting any deficiencies in the program might threaten the agency. "We must realize that our reason for existence lies only in how well we accomplish the ultimate objective of our agency: *the conservation of food, feed, and fiber essential for the survival of our nation,*" Fancher lectured the entomologists working for him.[79] So, beleaguered by turncoats and worried about the survival of

73. C. C. Fancher to E. D. Burgess, 3 May 1957, fire ant file, box 197; C. C. Fancher to E. D. Burgess, 29 November 1957, fire ant files (1957); L. J. Padget to E. D. Burgess, 11 March 1959, fire ant file, box 213; C. C. Fancher to E. D. Burgess, 18 January 1961, fire ant files (1961); C. C. Fancher to E. D. Burgess, 23 January 1961, fire ant files (1961), all in PPC papers.

74. The numbers are from C. C. Fancher to PPC supervisors in charge, 2 November 1960, fire ant file (1960), PPC papers.

75. C. C. Fancher to D. R. Shepherd, 31 May 1960, fire ant files (1960), PPC papers.

76. P. Daniel, "A Rogue Bureaucracy: The USDA Fire Ant Campaign of the late 1950s," *Agricultural History* 64 (1990): 109.

77. W. L. Popham to M. R. Clarkson, 1 July 1958, regulatory crops, 1958, fire ants file, box 943, General Correspondence, 1 (UD), ARS papers.

78. P. Daniel, "A Rogue Bureaucracy: The USDA Fire Ant Campaign of the late 1950s," *Agricultural History* 64 (1990): 112.

79. C. C. Fancher to Plant Pest Control Supervisors and Inspectors, Southern Region, 26 January 1961, information-speeches file, box 98, PPC papers. Emphasis in original.

their bureaucracy, they did not contemplate revising their eradication program, but instead struck back by mounting a public relations campaign. The counteroffensive was successful—the PPC maintained a great deal of public support, as well as support in both the federal government and among the states. But victory came with a cost. The USDA's critics had identified real problems: the insecticides were killing animals; the chemicals did contaminate cows' milk; support for the program was not universal. As the program continued, these problems piled up and eventually brought the eradication campaign to a halt.

The beginning of the end came in late 1959, when the Food and Drug Administration (FDA) outlawed the presence of any heptachlor residues in human food.[80] The public health community intervened very little in the controversy over insecticides, and when they did it was usually to support the use of chemicals, since insecticides controlled disease-bearing insects such as mosquitoes.[81] By 1959, for example, the public health official assigned to monitor the fire ant program had done no research, while a public health journal had praised the fire ant program for ridding the nation of an insect scourge.[82] But, despite this apathy, the FDA occasionally did enter the fray—warning consumers that cranberries might be contaminated with a dangerous insecticide in 1959, for example—and the PPC entomologists exposed their program to this kind of sudden reversal by ignoring the warnings of Edward F. Knipling, head of the USDA's Entomological Research Division (see chapter 2), and the complaints of wildlife biologists.[83] If PPC entomologists had worked diligently to minimize the danger done by heptachlor, rather than insisting that it posed little threat, the FDA might have cooperated with the PPC to curb, but not outlaw, the insecticide's use. Instead, when the PPC's position hardened, the FDA restricted the use of heptachlor, essentially banning the insecticide's use on vegetable crops and severely limiting its use on pastures.

The PPC put on a brave face, but this problem was close to insurmountable. Agricultural officials tapped Wayland Hayes of the U.S. Public Health Service, a proponent of pesticides, to explain the changes to

80. E. D. Burgess to B. T. Shaw, 18 February 1960, fire ant files (1960), PPC papers.

81. T. R. Dunlap, *DDT: Scientists, Citizens, and Public Policy* (Princeton: Princeton University Press, 1981), 39–55.

82. On the failure of the public health community to monitor the fire ant program, see "Minutes of the Southern Regional Fire Ant Research Conference," 15–16 December 1959, carton 4, series 1393, R. A. Gray Building, Florida State Archives (hereafter FL papers). The article highlighting the need to control the ant is F. G. Favorite, "The Imported Fire Ant," *Public Health Reports* 73 (1958): 445–48.

83. On the FDA, see C. J. Bosso, *Pesticides and Politics: The Life Cycle of a Public Issue* (Pittsburgh: University of Pittsburgh Press, 1987), 71–78, 96–100.

nervous field entomologists. He told them that no new information about heptachlor's toxicity had been found—it remained as safe as ever—but that different rules now governed when it could be applied.[84] E. D. Burgess, head of the PPC, hoped that as much as 95 percent of infected areas could still be treated. Hay pastures were immune from the new regulations, since humans did not eat hay, and, by splitting the two-pound treatment into two quarter-pound treatments, and then holding cattle off pasture lands for two weeks, both Burgess and Fancher thought that they could still spray ranches without contaminating cows' milk, although Fancher admitted that there was no research to back this approach.[85] But, their optimism did not bear out. An experiment at LSU showed that even ninety days after the PPC sprayed heptachlor on a cow pasture, a small amount of residue could be found in milk.[86] LSU's Leo Newsom said he would caution farmers against using any heptachlor on pastures.[87] In January 1961, Fancher whined, "About all we are doing at present is marking time; therefore our over-head cost is continuing without killing ants."[88]

The FDA's decision coincided with Alabama's and Florida's opting out of the program, making it increasingly difficult to balance the funding for the federal eradication program.[89] Many people throughout the South hated the ant, but they did not want to pay to eradicate it. The program needed to be reassessed. On February 9, 1961, Popham suggested that the campaign should be overhauled, and its goal redirected from eradication to control. The PPC would concentrate on keeping the ant from spreading, but "the eradication of the pest within areas of general infestation is not feasible," he admitted.[90] Fancher agreed.[91]

The Perfect Pesticide

Popham's admission seemed to lay to rest any hopes of eradicating the ant. But it was out of the wreckage of this first program that the seeds for

84. Hayes's comments are noted in C. N. Smith to E. F. Knipling, 21 December 1959, pesticide residue research lab-7-residues file, box 2, Entomology Research Division: Entomology Director's Correspondence, 1959–1965, 1055 (A1), ARS papers. For Hayes's ideas more generally, see W. J. Hayes, "Pesticides in Relation to Public Health," *Annual Review of Entomology* 5 (1960): 379–404.

85. On Burgess's advocacy of this approach, see E. D. Burgess to B. T. Shaw, 18 February 1960, fire ant files (1960), PPC papers. For Fancher's acknowledgment that research was lacking, see C. C. Fancher to State Supervisors, 14 March 1960, fire ant files (1960), PPC papers.

86. L. J. Padget to C. C. Fancher, 26 January 1961, fire ant files (1961), PPC papers.

87. Ibid.

88. C. C. Fancher to E. D. Burgess, 3 January 1961, fire ant files (1961), PPC papers.

89. C. C. Fancher to E. D. Burgess, 18 January 1961, fire ant files (1961), PPC papers.

90. W. L. Popham to B. T. Shaw, 9 February 1961, fire ant files (1961), PPC papers.

91. C. C. Fancher to E. D. Burgess, 9 March 1961, fire ant files (1961), PPC papers.

a second eradication program would sprout. When in 1957 the USDA initially announced its intention to wipe the imported fire ant from the face of the country, the PPC opened a laboratory in Gulfport, Mississippi to support the program. The Entomology Research Division's Clifford Lofgren was chosen to run the lab, with technical support from others in the ERD.[92] He was supposed to follow the research outlined by Burgess and look for cheap and potent chemicals.[93]

Over the first few years of its operation, the laboratory proved beneficial. In 1958, Lofgren found that one-and-a-quarter pounds of heptachlor spread on each acre killed as efficiently as two pounds; the PPC shifted its protocol in response.[94] About a year later, Lofgren and his staff found that four ounces of heptachlor spread across a single acre twice in ninety days was just as effective.[95] The USDA again altered its protocol, using the less intense dose on areas where wildlife biologists looked for dead animals, hoping to deflect continued criticism. [96] But, while the research focused on chemicals—one entomologist assigned to the lab wanted to leave because there were not enough biological studies—Lofgren, with Fancher's support, pushed the laboratory toward biological investigations, studying the ant's natural history (although he never quantified the damage done by the ant, still assuming that E. O. Wilson's early work was correct) and expanding his experiments to include research on biological control.[97] Like Leyburn Lewis (see chapter 2) before him, he found a mold that seemed to attack the insect. It proved an ineffective weapon, however.[98] Lofgren

92. C. N. Smith to A. W. Lindquist, 30 October 1957, fire ant file, box 65, Entomology Research Division: General Correspondence, 1954–1958, 5 (UD), ARS papers.

93. E. D. Burgess to E. F. Knipling, 7 October 1957, fire ant files (1957), PPC papers.

94. C. S. Lofgren to C. N. Smith, 12 December 1958; C. S. Lofgren to A. W. Lindquist, 22 December 1958, both in fire ant file, box 30, Entomology Research Division: General Correspondence, 1954–1958, 5 (UD), ARS papers.

95. C. C. Fancher to State Supervisors, 14 March 1960, fire ant files (1960), PPC papers.

96. E. D. Burgess explained the plan to use the small dose of poisons on areas monitored by wildlife biologists in E. D. Burgess to W. E. Blasingame, 14 April 1961, fire ant files (1961), PPC papers.

97. On the studies being mostly chemical in nature, see C. S. Lofgren to E. O. Wilson (attached to C. S. Lofgren to E. D. Burgess, 10 February 1959), fire ant file, box 213, PPC papers; on the entomologist's dissatisfaction, see V. Adler to A. W. Lindquist, 8 July 1958, personnel file, box 63; on Fancher's support for biological research, see C. N. Smith to A. W. Lindquist, 30 October 1957, fire ant file, box 65, all in Entomology Research Division: General Correspondence, 1954–1958, 5 (UD), ARS papers. On Lofgren's accepting Wilson's conclusions about the economic impact of the ant without doing further research, see C. S. Lofgren to E. O. Wilson (attached to C. S. Lofgren to E. D. Burgess, 10 February 1959), fire ant file, box 213, PPC papers.

98. C. S. Lofgren to A. W. Lindquist, 1 April 1958, fire ant file, box 30, Entomology Research Division: General Correspondence, 1954–1958, 5 (UD), ARS papers.

also experimented with *Bacillus thuringiensis*, better known as B.t., a bacterium that kills a number of insects, but the pathogen did not work against the fire ant.[99] In 1961, a USDA chemist noted, "It is evident that the entomology section has followed Dr. Knipling's original program."[100]

In the early 1960s, as the eradication program slowed in response to the FDA's ruling effectively banning the use of heptachlor, Lofgren began to experiment with these other methods more intensively, eventually making a breakthrough that would reinvigorate the eradication effort. Along with researchers at Auburn, Lofgren and his staff turned their attention to baits: mixtures of poisons, attractants, and food.[101] Because the ant was supposed to eat the bait—and share it with its nestmates—only miniscule amounts of poison would be needed to control the insect, helping to protect wildlife and humans. Marion Smith and other entomologists had used baits against the Argentine ant, and as recently as the late 1950s Harvard's William Brown had criticized the USDA for not considering them as weapons against the imported fire ant. Brown noted Wilson's recent work showing that food collected by workers quickly reached the queen, and so a bait could be very effective.[102] Lofgren first looked at a poison called "kepone," which seemed effective until the USFWS's James DeWitt showed that it altered the biochemistry of pheasants, giving male birds female sexual traits.[103] Soon afterwards, a chemical known as "GC1283," later to be better known by its trade name "Mirex," captured Lofgren's attention. Appropriately enough, in the past the chemical had been used to make materials fire retardant. Lofgren found that mixing GC1283 with soybean oil as an attractant and grits to give it bulk made an effective bait that contained a remarkably small amount of poison, only 1.7 grams per acre in the first trials.[104] (One entomologist said that Mirex was

99. C. S. Lofgren to A. W. Lindquist, 22 December 1958, fire ant file, box 30, Entomology Research Division: General Correspondence, 1954–1958, 5 (UD), ARS papers.

100. G. H. Gaddis to K. Messenger, 2 March 1961, fire ant files (1961), PPC papers.

101. C. C. Fancher to State Supervisors, 14 March 1960, fire ant files (1960), PPC papers.

102. On Smith's work, see R. W. Harned and M. R. Smith, "Argentine Ant Control Campaign in Mississippi," *Journal of Economic Entomology* 15 (1922): 261–64. Brown's opinions are in W. L. Brown to W. C. McDuffie, 10 April 1957, fire ant file, box 65, Entomology Research Division: General Correspondence, 1954–1958, 5 (UD), ARS papers.

103. On the rise and fall of kepone, see D. R. Shepherd to C. C. Fancher, 3 March 1961, insecticides file, box 98; J. B. DeWitt to L. J. Padget, 31 January 1961, and F. S. Arant to L. J. Padget, 10 February 1961, both in fire ant files (1961), all in PPC papers.

104. On the chemistry of Mirex, see R. L. Metcalf, "A Brief History of Chemical Control of the Imported Fire Ant," in *Proceedings of the Symposium on the Imported Fire Ant,* ed. S. L. Battenfield (Washington, D.C.: Government Printing Office, 1982), 124–27.

"safer than soap"; it was 99.925 percent grits and soybean oil.)[105] Over the years, the scientists experimented with various formulations of Mirex and protocols for its use, eventually settling on a program that called for spraying the bait three times every twelve to eighteen months.[106] In tests, the bait killed over 99 percent of the ant colonies, failing in only three experiments, but Lofgren found that the chemicals applied in those areas were contaminated with an ant repellent.[107]

Cries of excitement greeted the discovery of Mirex. Lofgren and others at the laboratory received awards for their work.[108] Secretary of Agriculture Orville Freeman called the bait "the perfect pesticide."[109] Fancher quickly latched onto the new chemical, asking his superiors if he could start using it immediately, although, he admitted, "the information we have on the toxic bait (corn cob grits and soybean oil) is extremely limited and the results are far from being conclusive—and, we might add, not quite up to our wishes."[110] Agriculture officials in Washington readily approved the use of the bait,[111] And Mirex became the standard tool for controlling the red imported fire ant. Between July 1963 and April 1964, the USDA sprayed 33,584 acres with heptachlor, but 1,213,485 with Mirex.[112] Upon hearing of the new chemical, Georgia's commissioner of agriculture J. Phil Campbell resurrected his state's spraying program, removing control from William Blasingame, the state entomologist—possibly because he had not alerted Campbell of Mirex's existence quickly enough, Fancher thought—and replacing him with Jack Gilchrist, a former journalist.[113] Fancher hated the change (complaining that Gilchrist had "about as much knowledge of or experience in directing an eradication program as my young daughter"), but Campbell proved a dedicated proponent of the new bait.[114]

105. Quoted in J. Thompson, "Insecticide for Fire Ants," 11 October 1962, fire ant file, box 128, PPC papers.

106. M. K. Hinkle, "Impact of the Imported Fire Ant Control Programs on Wildlife and the Quality of the Environment," in *Proceedings of the Symposium on the Imported Fire Ant*, ed. S. L. Battenfield (Washington, D.C.: Government Printing Office, 1982), 131.

107. C. C. Fancher to E. D. Burgess, 22 November 1961, fire ant files (1961), PPC papers.

108. "USDA Honors ARS Individuals, Groups . . . for Superior Service," *Agricultural Research* 11 (June 1963): 3–4.

109. Quoted in J. L. Whitten, *That We May Live* (Princeton: Van Nostrand Press, 1966), 115.

110. C. C. Fancher to E. D. Burgess, 22 November 1961, fire ant files (1961), PPC papers.

111. L. G. Iverson to C. C. Fancher, 13 December 1961, fire ant files (1961), PPC papers.

112. N. C. Brady to A. Ribicoff, 6 April 1964, committees file, box 17, Entomology Research Division: Entomology Director's Correspondence, 1959–1965, 1055 (A1), ARS papers.

113. F. Stancil to C. C. Fancher, 17 December 1962; C. C. Fancher to E. D. Burgess, 17 December 1962, both in fire ant files, box 128, PPC papers.

114. Fancher's observation is in C. C. Fancher to E. D. Burgess, 17 December 1962, fire ant files, box 128, PPC papers.

Wildlife biologists also embraced Mirex. Georgia quail biologist Herbert Stoddard used it on his property, confident that it would not hurt his quail, and the USFWS's Daniel Janzen said that the service was "heartened" by "the development of the Mirex formulation, as all our tests to date give evidence of its being a product which will permit control of fire ants without adverse effects on wildlife."[115] The USFWS allowed it to be used on wildlife refuges.[116] Even the Jackson, Mississippi zoo allowed the USDA to spray Mirex over its grounds. The incidental control of the white-fringed beetle and other insects could no longer be accomplished, since Mirex only killed ants, but the USDA could spread the bait more liberally than heptachlor.

Despite these kudos, however, the eradication program continued to lose steam. John F. Kennedy's election in 1960 ushered in new agricultural officials who had less invested in the eradication of the insect. (His secretary of agriculture, Orville Freeman, called the program "a classic case of how not to do things.")[117] By Johnson's administration—in which Freeman was still secretary and when he made his disparaging remark— enthusiasm was even weaker, so when Johnson's budget office pressured executive agencies to curtail spending, agriculture officials sacrificed the imported fire ant campaign.[118] The Gulfport laboratory was shut down. Eradicating the ant was perceived as a "low priority program," and eliminating it would have "the least damaging effects on our agricultural production."[119] Although Congressman Whitten's committee restored the funding—$3.3 million for the 1966 fiscal year—the budget office still thought that the program was worthless. It pressed for an independent evaluation of the program, which the USDA asked the National Academy of Sciences to undertake.[120] An august body of scientists who aided the government in making scientifically related policy decisions, the academy organized a panel under Harlow B. Mills that included William Creighton, Leo Newsom, Maurice Baker, and several

---

115. H. L. Stoddard to C. C. Fancher, 30 June 1962, fire ant files, box 128; D. H. Janzen to B. T. Shaw, 31 October 1962, insecticides-toxic effects file, box 134. See also L. G. Iverson to C. C. Fancher, 5 September 1962, fire ant files, box 128; M. F. Baker, "New Fire Ant Bait," n.d., fire ant cooperation file, box 165; R. E. Lane to USDA, 3 December 1962, fire ant files, box 128, all in PPC papers.

116. "Fire Ant Killer Tested at Refuge," n.d., fire ant files, box 128, PPC papers.

117. The Secretary to George Mehren, 27 June 1967, box 4679, insects (June) file, General Correspondence, 1906–1975, PI 191 and 1001 A-E (UD), Secretary of Agriculture papers, RG 16, National Archives and Records Administration, College Park, Md. (hereafter SOA papers).

118. H. Wellford, *Sowing the Wind* (New York: Grossman, 1972), 292–94.

119. These were stock phrases. For an example, see N. C. Brady to S. O. Long, 29 March 1965, insects file, box 4309, SOA papers.

120. G. W. Irving to O. L. Freeman, 20 June 1967, insects (June) file, box 4679, SOA papers.

others unsympathetic to the eradication program.[121] (It was during the course of the panel's investigation that Creighton met Murray Blum.) The panel ruled that the program for eradication of the red imported fire ant could not successfully be realized.[122]

The eradication ideal, though, was proving to be as resilient as the imported fire ant itself, and even as the National Academy of Sciences dismissed the possibility of eradication, a new program took shape, fueled by the success of Mirex and two other developments, one technical, the other, more important one, a constellation of political events. The technological innovation came in the form of a newly developed electronic aviation guidance instrument called the "Decca system." During the first eradication program, the PPC had had trouble making certain that insecticide was evenly sprayed across the South. "Very frankly," Fancher had lectured his underlings in 1961, "some of the results have been terribly disappointing. . . . [C]overage of insecticides was far from being uniform, and oftentimes complete skips were found."[123] Eradication was impossible under such circumstances, he said, since the ant could survive on the untreated areas. Installing the Decca system in airplanes helped to increase the accuracy of insecticide applications and the likelihood of eradicating the ant.[124]

Then, in the late 1960s, a series of events reconfigured the political environment, making the Decca system and Mirex seem the perfect tools to use in another attempt to eradicate the red imported fire ant. In 1967, the state commissioners of agriculture lobbied to restart the program. The impetus for this renewed interest probably had more to do with politics than a resurgence of calls by the public, as political scientist Harrison Wellford argues.[125] Throughout the 1960s, the public

---

121. On the panel's work, see P. M. Boffey, *The Brain Bank of America: An Inquiry into the Politics of Science* (New York: McGraw-Hill, 1975), 211–14. Also see the files in the National Academy of Sciences archives, Washington, D.C. (hereafter NASc papers), and Creighton's collection of material in the William Steel Creighton papers, an unprocessed box of material incorporated in the uncatalogued E. O. Wilson papers, Library of Congress, Washington, D.C. (hereafter Creighton papers).

122. "Appraisal of Programs for Fire Ant Control," 1 March 1967, Committee on Fire Ants, NASc papers.

123. C. C. Fancher to PPC Supervisors and Inspectors, 26 January 1961, information file, box 98, PPC papers.

124. D. K. Henderson, "Decca System Looks Good for Large Scale Ag Operations," *American Aviation* 30 (1966): 30–34; "Fire on Fire Ants," *Agricultural Aviation* 11 (1969): 5, 8; C. D. Paget-Clarke, "Electronic Guidance System Used in Fire Ant Eradication Programmes," *Agricultural Aviation* 13 (1971): 89–90; B. M. Glancey, "Evaluation of an Electronic Guidance System for Aircraft for Bait Application in the Imported Fire Ant Program," *Agricultural Aviation* 15 (1973): 38–40.

125. H. Wellford, *Sowing the Wind* (New York: Grossman, 1972), 301–4.

support for an eradication program had remained soft.[126] But the eradication program guaranteed a steady stream of money into state coffers, money that could be important to political careers. Agricultural Commissioner Campbell, for example, reportedly diverted some of the federal money into a slush fund, possibly to bankroll his run for governor of Georgia; and in the late 1970s one unnamed federal official told the *Washington Post* that other agricultural commissioners used the money to "make jobs, launch fleets of spray planes and get up at the Fourth of July picnics and tell everyone that without us to get rid of those ants you would not be here."[127] But whatever the cause of the reinvigorated interest, the lobbying worked, and Florida senator Spessard Holland, matching Whitten's enthusiasm for the program, ordered the USDA to investigate another eradication campaign against the imported fire ant.[128]

Meanwhile, Richard Nixon's presidential campaign was remaking the political landscape. Nixon set out to win the South for the GOP in the 1968 election, breaking the Democrat's stronghold on the region.[129] The fire ant played a small part in Nixon's southern strategy. While campaigning in Georgia, the candidate promised that his administration would work to destroy the insect.[130] Then, after he won the election, Nixon appointed J. Phil Campbell Under Secretary of Agriculture in charge of insect control operations (fig. 13). Like most southerners, Campbell had Democratic ties, and not everyone was happy to see him join Nixon's cabinet, but I suspect that the choice was made to solidify the alliance that the president was trying to forge with the South.[131] Well aware that the ant could be used as a lever of political power—Campbell once said, "The last several years I was in Georgia I could have defeated any member of the Georgia legislature from the ant-infested region of the State by going to his home territory and telling his voters he had not cooperated in fighting fire ants"—Campbell worked to dispel the USDA's reluctance to mount

126. Ibid., 300–1.

127. On Campbell, see ibid., 301–4. The quotation is from B. Richards, "EPA Ponders Approving a New Weapon against Fire Ant," *Washington Post,* 14 February 1978, A9.

128. G. W. Irving to O. L. Freeman, 20 June 1967, insects (June) file, box 4679, SOA papers.

129. D. T. Carter, *George Wallace, Richard Nixon, and the Transformation of American Politics* (Waco, Tex.: Markham Press Fund, 1992).

130. A. Wolff, "Big Schemes for Little Ants," *Audubon* 73 (March 1971): 122.

131. For disapproval of the appointment, see F. LeRoux to P. Belcher, 17 January 1969; C. J. Brown to B. N. Harlow, both in Department of Agriculture [1 of 3] file, box 3, John Whitaker subject files, White House Central Files, Richard M. Nixon papers, National Archives and Records Administration, College Park, Md. (hereafter Nixon papers).

Figure 13. Under Secretary of Agriculture J. Phil Campbell led the battle against the ant in Georgia, and then helped to reinvigorate the USDA's program to eradicate the insect when he joined that agency. (From the National Archives and Records Administration, negative 16-BN-33358.)

another ant eradication effort and steeled the agency for a new attack on the insect.[132]

Campbell found an ally in Mississippi's commissioner of agriculture Jim Buck Ross. Elected in 1968 for what would become a twenty-seven-year term of office, Ross brought catfish farming to the state, promoted the poultry industry, and called for a "coordinated southwide eradication program" against *Solenopsis invicta* (fig. 14).[133] Throughout 1969 and into the 1970s, J. Phil Campbell and Jim Buck Ross beat plowshares into swords, calling the insect "one of the most vicious and destructive pests ever to enter the U.S. from a foreign country," but emphasizing different aspects of the insect than before.[134] While in past years the ant had been considered an agricultural pest, Campbell now argued

132. Quotation from J. P. Campbell to J. C. Whitaker, 23 August 1971, insects July 26 to Oct. 31 file, box 5407, General Correspondence, 1906–1975, PI 191 and 1001 A-E (UD), SOA papers.

133. Ross's biographical data is in "Jim Buck Ross, Power in Miss. Agriculture, Dies," 15 December 1999, *Memphis Commercial Appeal*, A16.

134. J. B. Ross, "Speech," 29 July 1969, General Information to 1970 file, series 2012: fire ant correspondence, Mississippi Department of Agriculture and Commerce papers, Mississippi Department of Archives and History, Jackson, Miss. (hereafter MDAC papers).

Figure 14. The imported fire ant, as rendered by the Mississippi Department of Agriculture and Commerce. (From general info to 1970 file, Mississippi Department of Agriculture and Commerce papers, Mississippi Department of Archives and History.)

that it was a "people pest."[135] Ross canvassed Mississippi farmers for tales of the ant attacking people, feeding whatever stories he could elicit to Campbell and the press. One release, for example, quoted Ross as saying, "Almost every week, we learn of some new tragedy involving an attack by fire ants," and detailing how parents thought that their daughter had died when fire ants swarmed her after a car accident.[136]

This reconstruction of the ant resulted, I think, both from the South's changing demographics and from political demagoguery. As farms industrialized and more people moved to cities and suburbs, they less often confronted the ant in agricultural fields and more often felt its sting in their yards or in neighborhood parks. Thus, the ant seemed a menace to the homeowner and outdoor enthusiast more than a scourge of the southern farmlands. In this respect, the claims about the insect's threat propounded by Campbell and Ross reflected the actual experiences of many southerners. But they also stretched the truth about the ant. Some doctors responding to Ross's call for reports of ant attacks, for example, told him that the insect was not a major hazard

135. See "Opening Statement of Georgia Commissioner of Agriculture Phil Campbell before National Academy of Science Committee for Study of the Imported Fire Ant," 21 August 1967, Creighton papers.

136. News from MDAC, n.d., insects Jan. 21 to June 20 file, box 5407, General Correspondence, 1906–1975, PI 191 and 1001 A-E (UD), SOA papers.

to people. Ross, though, only passed on to the USDA those responses representing "the good side (our side)."[137] He also tried to argue that the ant's venom was a "major pollutant."[138] Ross made these exaggerations, I suspect, in response to another consequence of the changing demography from rural to urban and suburban: as more people moved to the city and suburbs, agricultural institutions lost their power. (Nixon even considered getting rid of the USDA.) By making the ant into a horror, and by positioning their administrative bureaucracies as necessary for effectively dealing with the threat, Campbell and Ross sought to make agricultural entomology a critical factor in the modernizing South.

Regardless of the validity of their arguments, Ross and Campbell pulled together the necessary ingredients for launching another ant eradication effort. By 1969, Ross convinced all the infested states to join the new attempt and hired C. C. Fancher to run his state's program.[139] And, with Campbell in charge, the USDA reopened the Gulfport laboratory. Lofgren published a paper arguing that the Decca system and Mirex made the eradication of the ant theoretically possible while another entomologist at the laboratory, William A. Banks, proved the bait's effectiveness in three separate field tests.[140] Familiar objections arose: Knipling, for example, argued that not enough research had been done to justify the new program, a point echoed by others.[141] But Campbell brushed these criticisms aside. He was confident in the support of the South's agricultural commissioners and the power of Mirex. In 1970 the USDA announced that it would undertake a twelve-year program to eliminate the ant from all 120 million acres that it occupied (fig. 15).[142]

137. L. Bailey to C. C. Fancher, 1 October 1970, correspondence 1969–1970 file, MDAC papers. This folder also contains the department's correspondence with Mississippi doctors.

138. J. B. Ross to W. E. Brock, 9 September 1970, insects Sept. 1 to Oct. 31 file, box 5227, General Correspondence, 1906–1975, PI 191 and 1001 A-E (UD), SOA papers.

139. On Ross's convincing the southern states to join the program, see "Never Before Has There Been a Southwide Coordinated Program to Eradicate the Imported Fire Ant," n.d.; Fancher's hiring is from "Nation's Top Expert on Fire Ants Heads Control Program," 23 July 1970, both in general information to 1970 file, MDAC papers.

140. C. S. Lofgren and D. E. Weidhaas, "On the Eradication of Imported Fire Ants: A Theoretical Appraisal," *Bulletin of the Entomological Society of America* 18 (1972): 17–20. On Banks's work, see W. A. Banks, B. M. Glancey, C. E. Stringer, D. P. Jouvenaz, C. S. Lofgren, and D. E. Weidhaas, "Imported Fire Ants: Eradication Trials with Mirex Bait," *Journal of Economic Entomology* 66 (1973): 785–89.

141. E. F. Knipling to D. R. Shepherd, 4 September 1970; J. H. Anderson to J. B. Ross, 21 July 1970, both in correspondence with Mississippi State University file, MDAC papers.

142. W. M. Blair, "Suit Seeks Curb on Insecticide Used to Fight Fire Ants in South," *New York Times*, 6 August 1970, 30.

Figure 15. "Invasion of the Stinging Fire Ant!" This Georgia advertisement captured all of the fears evoked by *Solenopsis invicta*'s invasion: the ant was martial threat that could only be met with belligerence. It attacked wildlife and interfered with farming. The ad also highlighted the danger that the insect posed to children—the imported fire ant was a people pest as well as an agricultural one. The ad made clear that there was no question but that the ant should be eradicated. (From fire ant file, box 98, Plant Pest Control Division papers.)

Perfect Pesticide No More

By the late 1960s, on the eve of the new fire ant eradication program, the environmental movement had grown in strength and power, shouldering aside—but not completely replacing—conservation allies and foes. Restructuring itself to accommodate the growing environmental ethic, membership numbers of the National Audubon Society increased from 32,000 in 1960 to 148,000 in 1970. (Within another decade, it reached 400,000.)[143] Environmentalists lobbied to protect the countryside, to limit pollution, and to ban the use of insecticides, especially DDT, which had been the target of much criticism since the Gypsy moth trial.[144] As historian J. Brooks Flippen notes, in its initial years, the Nixon administration tried hard to accommodate the environmental movement.[145] In the interest of winning the allegiance of a motivated bloc of voters, Nixon created the Environmental Protection Agency (EPA) and signed into law a number of legislative proposals regarding the environment. It was as they made the transition from political outsiders to insiders that environmentalists attacked the new eradication program.[146] Why? What had changed to make the perfect pesticide the object of environmental scorn? How did the environmentalists use their new power? And how did the transition from insider to outsider affect their attack on the Mirex program?

A few years after Mirex's introduction a familiar refrain was sounded, racing through professional circles and American popular culture: the chemical threatened wildlife and human health. This time, however, the animals in danger and the path of the toxin into the human body were different. Trace amounts of the chemical, biologists found, killed shrimp and crabs, weakening the base of the sea's food chain and endangering the livelihood of those who harvested the oceans' arthropods.[147] Mirex, some worried, also seemed to accumulate

143. M. E. Kraft, "U.S. Environmental Policy and Politics: From the 1960s to the 1990s," *Journal of Policy History* 12 (2000): 23.

144. On the environmental agenda, see S. P. Hays, *Beauty, Health, and Permanence: Environmental Politics in the United States, 1955–1985* (New York: Cambridge University Press, 1987). On DDT, see T. R. Dunlap, *DDT: Scientists, Citizens, and Public Policy* (Princeton: Princeton University Press, 1981).

145. J. B. Flippen, *Nixon and the Environment* (Albuquerque: University of New Mexico Press, 2000).

146. For a consideration of how outsiders win political power that is relevant to the growth of the environmental movement, see J. A. Morone, *The Democratic Wish: Popular Participation and the Limits of American Government* (New Haven: Yale University Press, [1990] 1998).

147. For overviews, see K. L. E. Kaiser, "The Rise and Fall of Mirex," *Environmental Science and Technology* 12 (1978): 520–28; R. L. Metcalf, "A Brief History of Chemical Control of the Imported Fire Ant," in *Proceedings of the Symposium on the Imported Fire Ant*, ed. S. L. Battenfield (Washington, D.C.: Government Printing Office, 1982), 124–27.

in the body of catfish, perhaps then passing to humans who consumed the fish.[148] In the late 1960s, some Mississippi State University (MSU) faculty banded into the Committee for Leaving the Environment of America Natural (CLEAN), in large part to protest the use of Mirex after evidence of its harmfulness appeared.[149] MSU biologist Denzel Ferguson led the campaign against Mirex, echoing arguments that had been made years before in earlier eradication programs. The ant, he said, was hardly a pest, while the bait used to kill it destroyed wildlife and threatened humans.[150] "There will be all manner of hell raised by the catfish farmers, and rightly so, if their product is condemned by the Pure Food and Drug People," Louis N. Wise, vice president of MSU and CLEAN member, wrote the USDA.[151] Moreover, as use of the chemical continued, evidence began to accumulate that Mirex harmed endangered birds, and a wildlife biologist said that the way Mirex moved through the food chain reminded him of DDT.[152] Mirex was starting to look far from perfect. In the face of the mounting evidence of Mirex's harmfulness, the Department of Interior reversed itself and banned the bait from all wildlife refuges.[153] Meanwhile, a study by the National Cancer Institute demonstrated that Mirex caused tumors in laboratory rodents.[154] Far from the perfect pesticide, Mirex, like its predecessors, killed wildlife and threatened human health. "We can't be quite so sure

148. "Doom Predicted for State Catfish," *Memphis Commercial Appeal*, 25 June 1970, general information to 1970 file, MDAC papers.

149. On CLEAN, see George P. Markin to Murray S. Bloom, 22 June 1972, Murray S. Blum papers, in author's possession (hereafter Blum papers); J. K. Stine, "Environmental Politics in the American South: The Fight over the Tennessee-Tombigbee Waterway," *Environmental History Review* 15 (1991): 1–24.

150. "Doom Predicted for State Catfish," *Memphis Commercial Appeal*, 25 June 1970, general information to 1970 file; "Mississippi Scientists Head Drive to Ban Widespread Mirex Campaign," [newspaper and date unknown]; "Anti-Pollution Group Seeks Ban of Mirex in Fight against Fire Ant," [newspaper and date unknown,] both in news clippings, 1970–1971 file; D. E. Ferguson, "Fire Ant: Whose Pest?" *Science* 169 (1970): 630. See also George P. Markin to Murray S. Blum, 22 June 1972, Blum papers.

151. L. N. Wise to N. D. Bayley, 29 May 1970, box 5227, insects May 1 to Aug. 31 file [2 of 2], SOA papers.

152. On birds, see R. L. Metcalf, "A Brief History of Chemical Control of the Imported Fire Ant," in *Proceedings of the Symposium on the Imported Fire Ant*, ed. S. L. Battenfield (Washington, D.C.: Government Printing Office, 1982), 125. The biologist's statement is from "Fire Ant Eradication Plan Meets Growing Opposition," September 1970, pesticides-1970 file [2 of 3], box 90, John Whitaker Subject Files, Nixon papers.

153. "Secretary Hickel Bans Use of 16 Pesticides on Any Interior Lands or Programs," 18 June 1970, pesticides-1970 file [2 of 3], box 90, John Whitaker Subject Files, Nixon papers.

154. J. R. Innes, B. M. Ulland, M. G. Valerio, L. Petrucelli, L. Fishbein, E. R. Hart, A. J. Pallotta, R. R. Bates, H. L. Falk, J. J. Gart, M. Klein, I. Mitchell, and J. Peters, "Bioassay of Pesticides and Industrial Chemicals for Mutagenicity in Mice: A Preliminary Note," *Journal of the National Cancer Institute* 42 (1969): 1101–14.

about the witch's brew of pollution that we are spreading around," warned one USDA opponent. "We may be taking one more step toward 'eradicating' a lot more than the fire ant."[155]

Thousands of letters opposing the program flowed into the White House and the USDA's Washington, D.C. office. The correspondence reflected the same themes that had motivated opposition to the fire ant wars since the 1950s. This latest fire ant eradication program was seen as just another chance for the USDA to increase its power and line the pockets of chemical industry allies while stomping on the liberties of average Americans. From Missouri: "Abandon this assault against Nature and turn your attention, instead, toward projects which will lessen, rather than add to, poisons which threaten all life on earth." From Nebraska: "It makes one wonder just who our real enemies are." From New York: "A local basis of spraying individual niches would be much more effective and safer." Letters came from across the South, from Ohio, Illinois, Pennsylvania, Colorado, Virginia, Utah, Connecticut, Arizona, California, Minnesota, Michigan, Maryland, and Massachusetts.[156] The *New York Times* editorialized against the program: "Why do rational men engage in this kind of senseless ecological gamble? The answer seems to be bureaucratic self-importance and political pressure. The more than $7 million which the Federal Government squanders on this program every year is spent by state agriculture departments and fattens their budget."[157] There were also what George P. Markin, head of the Gulfport Laboratory, described as a "rash of small court cases or legislative action aimed at banning Mirex in individual states," but these never succeeded.[158]

Ross and Campbell fought back, bullying their opponents. The Mississippi attorney general's office, after conferring with Ross, warned MSU's Denzel Ferguson, "You and your associates will be held responsible for any losses incurred by the state or any of its agencies, or any other parties, resulting from unwarranted interference with the state's program."[159] Ross also tried to turn public opinion against the biologist, drawing on sectional rhetoric that was especially poignant in light of

155. G. Laycock, "Eradicate Them One More Time," *Field and Stream*, December 1971, 130.

156. Letters in "Plant Protection Program: A Sample of Incoming Correspondence from the Public Relating to the Use of Mirex in the Imported Fire Ant Program, Received in the Fall and Spring of Fiscal Year 1971," ARS papers.

157. "Slow Learners," *New York Times*, 20 April 1971, Mirex—press releases file, box 73, EDF papers.

158. George P. Markin to Murray S. Blum, 22 June 1972, Blum papers.

159. R. H. Newcomb to D. E. Ferguson, 8 July 1970, general information to 1970 file, MDAC papers. See also R. Van den Bosch, *The Pesticide Conspiracy* (Garden City, N.Y.: Doubleday, 1978), 62.

the civil rights battles being waged at the time. "As is the case with a lot of northern-born and educated liberals," he wrote, "Mr. Ferguson has decided to step out of his own field (he is not an entomologist by education nor practice) and tell us what's wrong with us in Mississippi and the South. And, using the cloak of science and getting on the now popular 'anti-pesticide' and 'save the environment at any cost' bandwagon, he apparently hopes to build his case toward getting $10 million federal funds to do 'research' at the Mississippi Test Facility. If successful, he'd then be using State funds and facilities"[160]

The two eradication proponents also organized their own letter-writing campaign. Campbell told one of Ross's staff that the White House was "highly susceptible to outside pressures," and argued that an influx of mail from the South might influence administrative decision-makers.[161] He wanted ten thousand letters from each infested state.[162] It may be that Campbell received the flood of letters that he wanted, but the National Archives only preserves a few hundred letters in support of the program, almost all of them from school-aged children in Mississippi, and many of them filled with exaggerated stories. "I am for [the eradication program] continuing," one girl wrote, "because there are small children that ants can eat them." Another child asked that the program be continued because "I would rather put up with pollution, than [be] in a hospital unconscious because of an ant bite." A third reported, "One of my neighbors was found dead in a fire ant bed."[163]

While the wrangling between the USDA and its opponents continued, other environmentalists experimented with a new approach to bringing an end to the eradication program: suing the USDA. Convincing judges of the dangers of chemicals was, in many respects, easier and cheaper than breaking the iron triangle that protected pesticide use, and environmentalists used this strategy with increasing frequency in the attempt to ban a number of insecticides as well as control pollution.[164] The Long Island-based Environmental Defense Fund (EDF) made litigation its specialty. "Sue the bastards," was its war cry.[165] A grassroots group composed of lawyers and biologists, the EDF

160. Unmarked page, general information to 1970 file, MDAC papers.
161. R. Bailey to J. B. Ross, 21 December 1970, fire ant information file, MDAC papers.
162. Ibid.
163. Letters in "Plant Protection Program: A Sample of Incoming Correspondence from the Public Relating to the Use of Mirex in the Imported Fire Ant Program, Received in the Fall and Spring of Fiscal Year 1971," ARS papers.
164. A. MacIntyre, "Administrative Initiative and Theories of Implementation: Federal Pesticide Policy, 1970–1976," in *Public Policy and the Natural Environment*, ed. Helen M. Ingram and R. Kenneth Godwin (Greenwich, Conn.: JAI Press, 1985), 205–38.
165. On the EDF, see F. Graham Jr., *Since Silent Spring* (Boston: Houghton Mifflin, 1970), 251.

drew on the populist vocabulary honed by other USDA opponents and channeled it toward social change. Victor Yannacone, one of the EDF's founders, said, "[I]f there is a social need that must be filled, and something must be done for the People, with a capital P, there must be a legal way to do it." He and others with the organization were looking for that way. The legal system still had no provisions for people to sue for environmental concerns, but the EDF was intent to create such institutions. "We were practicing a form of legal guerilla warfare," Yannacone said.[166] Over time, the EDF and other environmental groups won the right to sue on behalf of the public welfare.[167]

In 1970, the organization, representing a collection of environmental groups, sued the USDA to stop the new fire ant campaign. Foreshadowing bigger battles to come, *EDF et al. v. Hardin* (Civil Case No. 2319-70) pitted those who wanted to eradicate the ant against those who saw the insect as merely a nuisance and Mirex as dangerous. William Brown, E. O. Wilson, CLEAN, and scientists from across the South lined up in support of the environmentalists' effort.[168] Building on the reformation of *Solenopsis invicta*'s reputation, Brown argued that while the insect occasionally could be a pest, it had also become part of the South's stable of ants. Its eradication—along with the unavoidable extermination of related ants that also ate the bait—could prove incalculably expensive. "We cannot say exactly what the result would be," Brown told the EDF, "except that it would be radical and pervasive in the affected area, therefore probably bad."[169] The EDF mocked the USDA's entomologists, asking "What do they expect to do—follow up the eradication program with an indigenous ant-planting program?"[170] But the death of an ant vital to the South's environment was not the only baleful result of the USDA's campaign. The insecticides also destroyed marine life and threatened endangered woodpeckers. Wilson called the program "The Vietnam of Entomology."[171] It seemed a fitting

166. Quotations from T. R. Dunlap, *DDT: Scientists, Citizens, and Public Policy* (Princeton: Princeton University Press, 1981), 148.

167. R. N. L. Andrews, *Managing the Environment, Managing Ourselves: A History of American Environmental Policy* (New Haven: Yale University Press, 1999), 239–42.

168. For a chronology of the case, see George P. Markin to Murray S. Blum, 22 June 1972, Blum papers.

169. Affidavit of William L. Brown, n.d, Mirex, EDF v. Hardin, No. 2319-70(1) file, box 69, Environmental Defense Fund papers, ms. 232, State University of New York, Stony Brook, N.Y. (hereafter EDF papers).

170. E. L. Rogers to W. L. Brown, 1 December 1970, correspondence (up to July 1, 1971) file, box 68, EDF papers.

171. Quoted in E. O. Wilson, *Naturalist* (Washington, D.C.: Island University Press, 1994), 117. See also "Fire Ant War Lost by U.S.," [newspaper unknown,] 29 September 1975, Mirex press

epitaph: a war undertaken to protect America from its enemy had devolved into a costly battle to achieve unclear goals lacking broad support.

Before a final judgment in the case could be rendered, however, political maneuvering by the Nixon administration took over. Observers at 1600 Pennsylvania Avenue had watched nervously as the trial unfolded. Nixon had promised the South that his administration would eradicate the ants and his aides were wary of alienating voters in the region by suddenly surrendering to pressure from northern liberals. But by 1971, only Georgia and Mississippi provided matching funds for the eradication effort, and Georgia's participation seemed in jeopardy as Governor Jimmy Carter reconsidered his support for the program.[172] Moreover, the surgeon general's office uncovered evidence that Mirex that accumulated in fish and shellfish could be harmful to humans.[173] In view of the circumstances, John C. Whitaker, Nixon's chief environmental advisor, concluded that the USDA would lose the court case.[174] Nixon, then, had two choices: he could concede the inevitable, but still appear to be the one in charge by ending the fire ant eradication program preemptively—and thus securing the votes of environmentalists. Or he could let the case continue, watch the USDA lose, but then "wink" at southern farmers and say that the administration had tried its best.[175] As it turned out, the administration tried to split the difference. Whitaker told the secretary of agriculture to limit the program—but not stop it—and soon thereafter the USDA announced that eradication was no longer the goal.[176] Government entomologists would only try to control the insect, and they would do so using far less Mirex than was needed to eradicate the bug.[177]

releases file, box 73, EDF papers; "Fire Ants: Vietnam of Entomology," [newspaper and date unknown,] news clippings, MDAC papers.

172. D. B. Rice to J. C. Whitaker, 8 March 1971, pesticides—1971 file [2 of 3], box 91, John Whitaker Subject Files, Nixon papers.

173. J. L. Steinfeld to J. C. Whitaker, 4 February 1971, pesticides—1971 file [2 of 3], box 91, John Whitaker Subject Files, Nixon papers.

174. J. C. Whitaker to J. D. Ehrlichman, 10 March 1971, pesticides—1971 file [2 of 3], box 91, John Whitaker Subject Files, Nixon papers.

175. Ibid. On Nixon's pesticide strategy more broadly, see J. B. Flippen, "Pests, Pollution, and Politics: The Nixon Administration's Pesticide Policy," *Agricultural History* 71 (1997): 442–56.

176. J. C. Whitaker to J. D. Ehrlichman, 10 March 1971, pesticides—1971 file [2 of 3], box 91, John Whitaker Subject Files, Nixon papers.

177. "Fire Ant Program, Department of Agriculture," 2 March 1971, pesticides—1971 file [2 of 3], box 91, John Whitaker Subject Files, Nixon papers; "Benefit/Cost Analysis Imported Fire Ant," Plant Protection Programs, USDA, March 1972, unfiled material, box 68, EDF papers.

Any celebration over what was at least a partial victory for the environmentalists, though, quickly dissipated. While the case progressed, the EPA convened a committee of scientists to review the evidence against Mirex. Led by C. H. van Middelem, a food scientist and apologist for insecticide use, the advisory panel also included Leo Newsom and J. R. Innes, the scientist who had overseen the tests that showed Mirex's carcinogenic potential. Admitting that more research was still needed before a final determination could be reached about the insecticide's safety and the ant's threat, the task force concluded that the use of Mirex to control the imported fire ant should be allowed to continue as long as the chemical was not applied to aquatic environments, wildlife refuges, or heavily forested areas. Most investigations, the panel concluded, "have failed to demonstrate any significant changes" in wildlife populations, while confirming that Mirex effectively controlled the "aggressive" ant.[178] Combined with the USDA's decision to retrench its operations, the panel's recommendation satisfied the judge—and the EPA—that the Department of Agriculture could continue to spread Mirex across the South. Eradication may have no longer been the goal; but, as environmentalists saw the situation, a potentially carcinogenic poison continued to rain down upon the South to control an insect that was merely a nuisance.

Mirex on Trial

Although the EPA and the EDF found themselves on opposite sides of the fire ant wars, the federal agency and the grassroots organization had otherwise forged a close relationship. Both were new, unsure of their strength, and hemmed in by other bureaucracies that had well-entrenched powers. Through a complex dance, however, the environmentalists and EPA staffers found that they could change pesticide policy. For their part, the EPA's officials tried to help environmental groups. The journal *Science* explained, "Apparently, the EPA is attempting to build itself a constituency to countervail pressures from the businesses it regulates. The effort is significant because if the EPA takes a neutral, adjudicative stance in disputes between environmentalists and polluting industries—as many believes it does—environmentalists are typically overwhelmed by the economic and political firepower of their adversaries."[179] The EPA, though, could not always lead. The same congressional leaders who supported the USDA's eradication campaigns

178. "Report of the Mirex Advisory Committee," 1 March 1972, pesticides May file, box 5604, SOA papers. See also "EPA Panel Finds Mirex 'No Serious Threat to Humans,'" 10 February 1972, Mirex press releases file, box 73, EDF papers.

179. "Toxic Substances: EPA and OSHA Are Reluctant Regulators," *Science* 203 (1979): 29–30.

controlled the environmental agency's budget. If the EPA moved too quickly, or with too much enthusiasm, Congress could draw the purse strings tight. In these cases, environmentalists found that they could prod the agency into motion by suing. The EPA could then explain to Congress that it had been forced into action—the environmentalists being a convenient scapegoat that allowed the agency to take positions that might have been untenable otherwise.[180]

As allies, the EPA and the EDF focused most of their attention on banning pesticides, especially those that posed a threat to humans, paying less attention to the chemicals' effects on wildlife than their cancer-causing properties.[181] The EPA lacked the authority to change the land-use practices that allowed the ant to spread and necessitated the use of insecticides; it lacked the resources to establish standards for tolerable levels of pollutants in the environment; and it lacked the desire to encourage alternative methods of control. In fact, it prevented their introduction. In the early 1970s, the California biotechnology company Zoecon pioneered a method of controlling the fire ant using an insect-growth regulator—a hormone analogue that prevented larvae from maturing into adults.[182] The EPA sensibly required that the chemical pass the same battery of tests as any insecticide had to do before it was approved for use. But when Zoecon asked the agency for a grant to help pay for the cost of the testing, the agency balked.[183] Leo Newsom also complained that the EPA's requirements "brought to a virtual standstill" studies of alternative techniques.[184]

The EPA, though, was rich in lawyers, and so a "sue the bastards" approach played to its strength.[185] The agency had inherited enforcement staff from the parts of the Departments of Health, Education, and

180. On the relationship between the EPA and environmental groups, see R. N. L. Andrews, *Managing the Environment, Managing Ourselves: A History of American Environmental Policy* (New Haven: Yale University Press, 1999), 241.

181. E. P. Russell, "Lost among the Parts Per Billion: Ecological Protection at the United States Environmental Protection Agency, 1970–1993," *Environmental History* 2 (1997): 29–51.

182. On Zoecon, see G. Bylinsky, "Zoecon Turns Bugs against Themselves," *Fortune* 88 (August 1973): 94–103

183. C. Djerassi, C. Shih Coleman, and J. Diekman, "Operational and Policy Aspects of Future Insect Control Methods," 12 February 1974; C. Djerassi to R. Long, 12 November 1973, both in insects, April 19 to July 16 file, box 5855, SOA papers.

184. L. D. Newsom to R. W. Long, 14 March 1974, insects, April 19 to July 16 file, box 5855, SOA papers.

185. For more analyses of the EPA's choices, see M. K. Lansy, M. J. Roberts, M. K. Landy, and S. R. Thomas, *The Environmental Protection Agency: Asking the Wrong Questions* (New York: Oxford University Press, 1990); M. Winston, *Nature Wars: People vs. Pests* (Cambridge: Harvard University Press, 1997), 158–59; R. N. L. Andrews, *Managing the Environment, Managing Ourselves: A History of American Environmental Policy* (New Haven: Yale University Press, 1999), 316–18.

Welfare, Interior, and Agriculture that had been folded into the new agency at its creation. And it was led by William Ruckelshaus, a former deputy state attorney general assigned to the Indiana State Board of Health, who had made a name for himself in Republican circles prosecuting polluters and helping to draft the Indiana Air Pollution Act. Appointed to head the EPA, Ruckelshaus said that it was "important to demonstrate to the public that the government was capable of being responsive to their expressed concerns; namely, that we would do something about the environment. Therefore, it was important for us to *advocate* strong environmental compliance, back it up, and *do* it; to actually show we were willing to take on the large institutions in society which hadn't been paying much attention to the environment."[186]

In 1973, environmentalists urged the EPA to reconsider the Mirex question. "The Mirex imported fire ant control program needs a thorough benefit-cost analysis," the EDF's Lee Rogers explained. "The issues are far broader than whether Mirex is a carcinogen and whether the fire ant stings are painful and cause a small percentage of the stung medical problems. The areas that need to be explained are the economics of the program versus the alternatives that might be adopted to minimize the nuisance of the fire ants."[187] The EPA agreed and announced its intention to ban the bait, at the time the only way to trigger a full-scale review. The USDA objected, however, and the two sides prepared to present their cases before an administrative judge. The EPA made the argument for the environmentalists—with support from the EDF—laying out a three-pronged attack. First, a series of witnesses disputed the USDA's claim that the ant was much of a pest. Newsom testified, as did Buren, and Stoddard's protégé, Ed Komarek, all stating, to varying degrees, that while the ant could sometimes be a nuisance, it did not deserve its reputation as one of the South's worst pests.[188] Komarek said, referring to the quail plantations that he managed, "Our highest quail populations occur in . . . areas . . . where *Solenopsis invicta* was or is in greatest abundance. If fire ants were harmful to quail we could not

186. EPA, "U.S. EPA Oral History Interview 1, William D. Ruckelshaus," 202-K-92-0003 (Washington, D.C.: EPA, 1993). Emphasis in original. See also J. Quarles, *Cleaning Up America: An Insider's View of the Environmental Protection Agency* (Boston: Houghton Mifflin, 1976), xv; R. N. L. Andrews, *Managing the Environment, Managing Ourselves: A History of American Environmental Policy* (New Haven: Yale University Press, 1999), 229–31.

187. L. Rogers to C. Wurster, 29 May 1973, correspondence, internal file, box 71, EDF papers. See also W. A. Butler to Executive Committee and Executive Director, 11 May 1973, correspondence (from July 1, 1971) file, box 68, EDF papers.

188. "Statement of William Franklin Buren," 29 January 1974, Buren file, box 62; "Statement of Leo Dale Newsom," 10 April 1974, Newsom file, box 63, both in EDF papers.

have built our quail populations up to such high levels."[189] Walter Tschinkel, a young entomologist who had recently joined the faculty at Florida State University, echoed these assessments. "Based on my experience as an insect biologist and resident of a fire ant infested area," he testified, "in most situations, the fire ant is not a serious pest and should be ranked as no more than a nuisance." In some ways, the ant was even beneficial—a necessary part of the environment, as William Brown had suggested in the first trial. Tschinkel said, "Once established in an area, the fire ant tends to become the dominant ant and, because of its carnivorous habits, will simultaneously become the dominant predator of other insects. Its removal would thus mean the removal of a major insect predator and has been shown in some cases to have adverse effects."[190]

Second, the EPA's witnesses questioned the effectiveness and safety of Mirex. Tschinkel testified that in his studies, the queen did not usually receive the bait. In one experiment, 80 percent of the workers ate Mirex, but the queen ate none. "While this does not prove that the queen never receives oil," he said, "it indicate [sic] that the flow of oil within the colony is dependent on conditions." The USDA had no idea what those conditions were—nor did anyone—and so many of the ant colonies may have survived. USDA entomologists may have overlooked the possibility that even where the worker ants had died and the nest looked moribund, deep inside the mound the queen may have been producing another colony, prepared to take advantage of the newly disturbed area.[191] Others argued that while the poison may not kill *Solenopsis invicta* it harmed marine life.[192]

Third, the EPA pointed out the absurdity of the program. EPA economist Fred Arnold testified that if farmers were left to their own devices, they would spray only as much chemical as necessary. But insofar as the USDA subsidized the cost of the bait—paying for half the program—southerners had an incentive to use more than they needed.[193] The department encouraged the overuse of a potentially

189. "Statement of E. V. Komarek," 21 February 1974, Komarek file, box 62, EDF papers.

190. "Statement of Walter R. Tschinkel," 7 February 1974, Tschinkel file, box 61, EDF papers.

191. Ibid. See also "Public Hearing to Determine Whether or Not the Registration of Mirex Should be Cancelled or Amended," F.I.F.R.A. Docket no. 293, 28 February 1974; "Walter R. Tschinkel, Direct Testimony," 28 February 1974, both in Tschinkel 2/28/74 file, box 61, EDF papers.

192. See the testimony of Joe Bowden, Henry Enos, Gene Fuller, Alva Harris, Terry Hollister, Jeffrey Lincer, Robert Livingston, Norbert Page, Dudley Peeler, Bart Puma, and G. Bruce Wiersma, all in EDF papers.

193. "Statement of Fred T. Arnold," 10 April 1974, box 64, Arnold file; "Fred Arnold, Direct Testimony," 12 and 13 June 1974, summaries file, box 68; "Public Hearing to Determine

carcinogenic chemical—and this despite the availability of alternatives. Buren and John Diekman, Zoecon's vice president, both testified that new, less harmful methods of killing *Solenopsis invicta* were on the horizon.[194] Yet, the agency argued, the USDA insisted on using a dangerous and possibly ineffective method. What kind of public policy was that?

The environmentalists' presentation was long—the hearing transcript more than ten thousand pages—complicated, and marred by difficulty coordinating schedules. No one was sure exactly how the hearing should proceed. Was the matter under review legalistic or scientific? Were the two sides adversarial or collegial?[195] Under the weight of such ambiguity, the participants muddled along, and sometimes the proceedings got a little loopy. Guy C. Jackson, attorney for the Louisiana and Texas Departments of Agriculture, for example, suggested that the ant might "bite on command." Perhaps, he said, "They have intelligence and are going to take over the world."[196] By 1975, two years after the hearing began, the litigants were exhausted and began contemplating a compromise. The EDF knew that its case had holes. Lawyers defending Mirex had almost impeached Tschinkel's testimony since he had never published an article on the ant.[197] His claims were vulnerable. The only evidence that would be damning was proof that Mirex caused cancer, but there was none.[198] For its part, Allied Chemical, the producer of Mirex, made some money from the sale of the bait, but not much: $1 million dollars each year against total

Whether or Not the Registration of Mirex Should be Cancelled or Amended," F.I.F.R.A. Docket no. 293, 13 June 1974, transcript, Arnold 6/13/74 file, box 64, all in EDF papers. See also "Gerald Carlson, Direct testimony," unmarked file, box 68, EDF papers.

194. "Statement of William Franklin Buren," 29 January 1974, Buren file, box 62; "John Diekman, Direct Testimony," 21 February 1974, summaries file, box 68, both in EDF papers. See also "E. W. Cupp, Direct Testimony," 20 February 1974; "S. Bradleigh Vinson, Direct Testimony," 21 February 1974, both in summaries file, box 68, EDF papers.

195. For these questions, see "Public Hearing to Determine Whether or Not the Registration of Mirex Should be Cancelled or Amended," F.I.F.R.A. Docket no. 293, 11 July 1973, transcript page 18, prehearing conference July 11 transcript file, box 60; "Public Hearing to Determine Whether or Not the Registration of Mirex Should be Cancelled or Amended," F.I.F.R.A. Docket no. 293, 15 March 1974, transcript page 3897, Dillier file, box 62, both in EDF papers.

196. "Public Hearing to Determine Whether or Not the Registration of Mirex Should be Cancelled or Amended," F.I.F.R.A. Docket no. 293, 25 October 1974, transcript pages 9068–69, no testimony file, box 65, EDF papers.

197. For the attack on Tschinkel's credibility, see "Public Hearing to Determine Whether or Not the Registration of Mirex Should be Cancelled or Amended," F.I.F.R.A. Docket no. 293, 28 February 1974, transcript pages 2795–2920; "Walter R. Tschinkel, Direct Testimony," 28 February 1974, both in Tschinkel 2/28/74 file, box 61, EDF papers.

198. W. A. Butler to EDF Mirex Clients, 20 March 1975, settlement negotiations file, box 72, EDF papers.

revenues of $1.24 billion in 1971. The trial was becoming too expensive for the company.[199] To complicate matters further, the EPA had lost its lead counsel, and the new legal team placed "a low relative priority on the Mirex case given what they see as higher priorities in pesticide regulation of other poisons having more conclusive health evidence (heptachlor/chlordane)," according to William A. Butler, the lawyer heading the case against Mirex for the EDF.[200] Together, the EDF and EPA had won outright bans on DDT and had strong cases against other chemicals, but Mirex, the perfect pesticide, looked like it might survive another attack. Hoping to avoid having a ruling imposed upon them by the judge, the parties agreed in March 1975 that the EPA would allow the USDA to spray two Mirex treatments over two years, but never near coastal areas. Although Allied Chemical had been selling its product for use in a program that required three treatments in fifteen months—that was how the department's entomologists achieved a 99 percent kill rate in field trials—the chemical company agreed with the reduction.[201]

The Banning of Mirex

The seeming compromise between the USDA, on the one hand, and the EPA and EDF on the other, quickly fell apart, in the process revealing the changed political landscape. In April 1975 the USDA unilaterally canceled the fire ant program—no eradication, no control, no work on the fire ant at all.[202] In June, the House agriculture committee held hearings to try to patch over the differences between the department and the agency, and the legislators took the opportunity to severely criticize the EPA.[203] Perhaps feeling confident, Under Secretary Campbell then pulled his lawyers from the negotiations and, according to the EDF, called state agriculture commissioners and told them to oppose the settlement that earlier had been reached.[204] Secretary of

199. On Allied's fatigue, see W. A. Butler to EDF Mirex Clients, 20 March 1975, settlement negotiations file, box 72, EDF papers. On the company's financial situation, see D. Shapley, "Mirex and the Fire Ant: Declines in the Fortune of a 'Perfect' Pesticide," *Science* 172 (1971): 359.

200. W. A. Butler to EDF Mirex Clients, 20 March 1975, settlement negotiations file, box 72, EDF papers.

201. W. A. Butler to EDF Mirex Clients, 24 June 1975, correspondence internal file, box 71, EDF papers. See also U.S. Congress, Subcommittee on Department Operations, Investigations and Oversight, Committee on Agriculture, *Fire Ant Eradication Program*, 94-1, 1975, 31–32.

202. J. E. Brody, "Agriculture Department to Abandon Campaign against the Fire Ant," *New York Times*, 20 April 1975, 46.

203. U.S. Congress, Subcommittee on Department Operations, Investigations and Oversight, Committee on Agriculture, *Fire Ant Eradication Program*, 94-1, 1975.

204. W. A. Butler to EDF Mirex Clients, 15 July 1975, correspondence internal file, box 71, EDF papers.

Agriculture Earl Butz told the press, "Continuing restrictions placed on the pesticide have finally made the program unworkable."[205] Apparently, Campbell hoped that southerners would be so frustrated that they would rise up and demand that the full-fledged eradication program be renewed, proving the wisdom of the USDA's original stance and making environmentalists seem out of touch with the needs of the people. The EPA administrator complained to Butz that the USDA's outright canceling the program made the EPA look bad and pressed for resumed negotiations, but, as Butz recounted to Campbell, "I indicated to him the time for talk was past."[206]

Campbell, though, overestimated the strength of his position and misjudged the power of his opponents. The volley of letters that he expected never materialized; a few southerners raised their voice in protest, a few publications condemned the EPA, but popular sentiment never swung decisively toward the USDA. The rise of environmentalism had changed the political landscape: the EPA had successfully annexed some of the USDA's turf. In the wake of Campbell's announcement, Allied Chemical quit the trial and ceased the production of Mirex, costing the USDA its strongest ally, for Allied's lawyers had been doing much of the work at the trial. Campbell was livid, seeing in Allied's decision some secret plot between environmentalists and the chemical company. He told Georgia congressman Dawson Mathis, "The EPA and Allied Chemical . . . seem to have been working very closely together in the past three years."[207]

With Allied out, the USDA took control of the court case, and the trial resumed until Mississippi Commisioner of Agriculture Jim Buck Ross swooped in and salvaged what he could out of the wreckage of the negotiations. In May 1976, Allied sold its Mirex-manufacturing plant to the State of Mississippi for one dollar, alienating the USDA and shifting power in the matter to Ross.[208] The coordinated southwide effort to eradicate the imported fire ant entered a new phase. The Magnolia State planned a small-scale version of the earlier federal program, first spraying along its borders, surrounding the insect, then driving it into

205. Quotation from C. Norman, "The Little Nipper Who Cost the South a Fortune," *Nature* 255 (1975): 94.

206. For the EPA's complaints, see J. R. Quarles to J. P. Campbell, 4 April 1975, insects May 13 to 30 file, box 5983, SOA papers. The quotation is from E. L. Butz to J. P. Campbell, 1 April 1975, insects April 1 to May 12 file, box 5983, SOA papers.

207. J. P. Campbell to D. Mathis, 31 July 1975, pesticides, July 1 to August 31 file, box 6012, SOA papers.

208. "Mississippi to Make Pesticide," [newspaper unknown,] 12 May 1976, correspondence, internal file, box 71, EDF papers; "South Still Fights Its Longest War, Learns to Live with Surly Fire Ants," 28 June 1976, news reports file, MDAC papers.

the Gulf of Mexico.[209] Mississippi also sold the bait to other states. Complete eradication was increasingly unlikely, but the states could at least continue to use Mirex.

Mississippi's ascendancy as a force in the fire ant wars seemed to fulfill the prophecies of Cold War jeremiads. The rush to kill insects using chemical pesticides in defense of the country had ended by turning the national security state into a socialist bureaucracy. The press quickly picked up on this connection between pest control and Cold War fears. "In an apparently unique business venture by a state government, Mississippi has taken over the production and sale of the controversial pesticide Mirex," wrote one newspaper.[210] Less circumspect, the *Star Journal* from Gulfport called Mississippi's arrangement "Magnolia-Scented Socialism."[211] "Many Mississippians do not know that their state entered into private business," wrote Dennis Dollar. "That's right. Today, our state is using taxpayers' money to finance the operation of a pesticide manufacturing plant in Aberdeen."[212] Another article began, "A losing effort to conquer a tiny, unconquerable insect has brought a strange and costly species of socialism to the Deep South. Today, the final citadel of this socialism is to be found, of all places, in Mississippi, whose politicians for decades urged the voters to resist any hint of socialism wherever it seemed to be nibbling at the woodwork of free enterprise."[213] In some small way, the socialism that nurseryman J. Lloyd Abbot and others had worried about came into being.

Confirmed in their fears regarding Mirex's application, environmentalists pressed to stop the use of the bait, but despite their increasing political strength, they still found the path to victory arduous and the struggle "tangled enough to make Faulkner's ghost blink."[214] In 1976, after Mississippi began producing Mirex, environmental researchers finally found the evidence that they needed to ban Mirex. The bait, the EPA discovered, degraded into kepone, a chemical that was known to cause cancer, and a government survey found that 44 percent of

209. On Mississippi's plans, see C. C. Fancher to S. Montgomery, 17 February 1972, congressional delegation, 1971–1972 file, MDAC papers.

210. "Mississippi to Make Pesticide," [newspaper unknown,] 12 May 1976, correspondence, internal file, box 71, EDF papers.

211. "'Magnolia-Scented Socialism' Invades Mississippi," *Gulfport Journal*, n.d., news reports file, MDAC papers.

212. Ibid.

213. "Solenopsis the Unconquerable," [newspaper unknown,] July–August 1976, Mirex—press releases file, box 73, EDF papers.

214. Quotation from B. Richards, "EPA Ponders Approving a New Weapon against Fire Ant," *Washington Post*, 14 February 1978, A9.

Mississippians had Mirex stored in their body's fat.[215] There seemed little hope for the insecticide's defenders. Thus, Mississippi asked the EPA to place a ban on the use of Mirex that would go into effect in 1978, giving the state time to use its remaining stocks.[216]

But, before the environmentalists could claim victory, another battle flared up, although this time around the positions were reversed and it was the opponents of chemical insecticides who were on the defensive. Chemists for the State of Mississippi reformulated Mirex into another chemical, called "ferriamicide," and Jim Buck Ross pushed hard for the EPA to approve it as an emergency replacement for Mirex, enlisting U.S. senator James O. Eastland from Mississippi to coordinate a political campaign. The senator threatened to kill legislation favored by the EPA unless the agency approved the use of ferriamicide. He whipped up public sentiment as well: over ten thousand letters from southern farmers demanding the right to use ferriamicide poured into Washington, D.C. According to the *Washington Post,* "A shaken senior EPA official said that the campaign for the pesticide was the most intense to hit the agency in years."[217] Eventually, the EPA agreed to authorize ferriamicide's use, although Barbara Blum, head of the agency, admitted, "Legally, we probably can't approve it."[218] Calling ferriamicide the "son of Mirex," the EDF responded by bringing suit against the EPA, arguing that the chemical should be outlawed: it degraded into photomirex, a possible carcinogen that was ten to one hundred times more toxic than Mirex, five times more toxic than kepone.[219] The environmentalists found unexpected allies in the chemical company American Cyanamid and former Georgia congressman Dawson Mathis. American Cyanamid had developed another chemical (Amdro) to control the ant, one that had passed the battery of tests the EPA required of all new pesticides. But Amdro was more expensive than ferriamicide and would

215. "Human Tissue Shows Mirex," 16 July 1976; "Mirex Found in Bodies Fuels Pesticide Controversy," 8 July 1976, both in Mirex file, NAS papers; "Mirex Found in 44 per cent of Mississippians," 25 August 1976, Mirex–press releases file, box 73, EDF papers.

216. "State Asks EPA to Ban Mirex Use in Two Years," n.d.; "Mirex's Phase out," n.d., both in news reports file, MDAC papers; "Mississippi Stymied by Lack of Fire Ant Poison," 23 November 1976, Mirex–press releases file, box 73, EDF papers.

217. Quotations and the political maneuverings to restart the program are from B. Richards, "EPA Ponders Approving a New Weapon against Fire Ant," *Washington Post,* 14 February 1978, A9.

218. W. Sinclair, "Politics Wins Battle on Pesticide to Curb Fire Ants," *Washington Post,* 15 February 1979, A33.

219. Quotation from J. Oglethorpe, "Ecologist Vows to Continue Crusade," *Jackson Daily News,* 23 February 1979, A4. Ferriamicide's problems are described in W. Sinclair, "Politics Wins Battle on Pesticide to Curb Fire Ants," *Washington Post,* 15 February 1979, A33.

never find a place in the market so long as ferriamicide was legal. The company hired Mathis to lobby the EPA on its behalf.[220]

The fight over ferriamicide continued for five years, during which time the EPA was split on the matter. Finally, in 1983, the battle came to an end when the anti-ferriamicide faction within the agency overcame those bureaucrats who had supported the reformulated bait, and the agency banned the chemical.[221] The environmentalists had won political power, achieved Rachel Carson's dream, and ended that of the PPC. Nature, the environmentalists said, was a finely balanced network; synthetic chemicals disrupted the system, bringing chaos to an ordered system. The red ant, however, a biological entity, could find a place in nature's economy. Humans could, too, if they stopped fighting nature and learned to accommodate themselves to the rhythms of the natural world. The end of the fire ant wars marked a step forward, the environmentalists promised—a step away from what Rachel Carson had called the "Neanderthal age" of biology's and philosophy's understanding of nature and toward human co-existence within it.

The Changing Imported Fire Ant

As the political landscape changed and environmentalists gained power at the expense of the USDA, the imported fire ant was changing as well, undermining hopes for an easy reconciliation between humans and the natural world. In its South American homeland, the ant comes in two forms: monogynous and polygynous. The monogynous form, as the name implies, has only one queen, while the polygynous has many. And it is the polygynous colony that is the more destructive of the two. By the mid-1970s, the polygynous form had become the dominant kind of *Solenopsis invicta* in many areas, responsible for huge amounts of damage.[222] Historian Pete Daniel suggests that the barrage of insecticides caused the increase in polygyny, more evidence of the USDA's arrogance and the damage caused by altering nature without first studying it.[223] There is, however, no evidence to support his

220. W. Sinclair, "EPA Pressed to Approve New Fire Ant Pesticide," *Washington Post,* 20 February 1982, A6.

221. "EPA Reverses Decision on Fire-Ant Killer," *Washington Post,* 4 July 1983, A13.

222. B. M. Glancey, C. H. Craig, C. E. Stringer, and P. M. Bishop, "Multiple Fertile Queens in Colonies of the Imported Fire Ant, *Solenopsis invicta,*" *Journal of the Georgia Entomological Society* 8 (1973): 237–38. See also B. M. Glancey, J. C. Nickerson, D. Wojcik, J. Trager, W. A. Banks, and C. Adams, "The Increasing Incidence of the Polygynous Form of the Red Imported Fire Ant *Solenopsis invicta* (Hymenoptera: Formicidae) in Florida," *Florida Entomologist* 70 (1987): 400–2.

223. P. Daniel, *Lost Revolutions: The South in the 1950s* (Chapel Hill: University of North Carolina Press, 2000), 85. See also E. Tenner, *Why Things Bite Back: Technology and the Revenge of Unintended Consequences* (New York: Knopf, 1996), 212.

supposition. The rise of polygyny, I think, teaches a different lesson: that the ant was not as benign as many environmentalists believed.

The rising prominence of polygynous fire ants in the 1970s was probably not a response to Mirex but, as biologists Kenneth Ross and Laurent Keller argue, to the changing demography of the invaders.[224] Monogynous nests produce queens that are heavy, with large fat reserves. Monogynous queens produce eggs quickly.[225] These traits allow the monogynous queens to fly from their nests, colonize new areas, and rapidly reproduce. They are well adapted for pioneering. "However," Ross and Keller note, "the success of dispersing [monogyne] queens probably decreases as appropriate nesting sites become filled, and as predation and brood raiding by workers from existing colonies increase."[226] Under these conditions, polygynous colonies do better. Their queens are smaller than their monogynous relatives, with less fat in reserve and a slower reproductive development. They do not leave the nest on mating flights, but instead recruit workers from the mother colony and travel by walking to a nearby site suitable for establishing a new nest.[227] This lifestyle allows the polygynous colony to succeed in densely populated areas, protecting young queens from predation. Polygynous ants probably made the trip from South America in the first half of the twentieth century, but remained largely unnoticed so long as the environment favored the monogynous colonies. As the density of the fire ant population increased, though, the advantage shifted to the polygynous ants and they came to be the dominant variety.

The increasing prevalence of polygyny changed the dynamics of the insect's invasion. More and more ants could be packed into the same area and the red imported fire ant dominated areas even more completely than in the past. In the 1990s, some Texas' fields supported over four hundred imported fire ant mounds per acre.[228] With one colony budding from another, the polygynous colonies spread across

224. K. G. Ross and L. Keller, "Ecology and Evolution of Social Organization: Insights from Fire Ants and Other Highly Eusocial Insects," *Annual Review of Ecology and Systematics* 26 (1995): 636.

225. K. G. Ross and L. Keller, "Genetic Control of Social Organization in an Ant," *Proceedings of the National Academy of Sciences* 95 (1998): 14,232.

226. K. G. Ross and L. Keller, "Ecology and Evolution of Social Organization: Insights from Fire Ants and Other Highly Eusocial Insects," *Annual Review of Ecology and Systematics* 26 (1995): 636.

227. E. L. Vargo and S. D. Porter, "Colony Reproduction by Budding in the Polygyne Form of the Fire Ant, *Solenopsis invicta*," *Annals of the Entomological Society of America* 82 (1989): 307–13.

228. R. L. Brown, "Fire Ants: Buggy Battalions Beating Mere Mortals," *Atlanta Journal-Constitution*, 21 June 1992, M2.

the landscape, decimating other ants. Entomologists found that the prevalence of workers from other species decreased by as much as 70 percent and the diversity of ant species plummeted by up to 90 percent after the polygynous fire ant invaded a field.[229] "When they sweep into an area, they obliterate most native ants, plus a good many spiders, worms, and other invertebrates," wrote environmental journalist Robert S. Devine in 1998. "Scientists don't yet thoroughly understand the ramifications of losing all these little creatures, but pessimism is justified."[230] The environmentalists had stopped the use of insecticides, but they did not stop the ant. In 1983, with all the chemicals devised to eradicate the ant banned, Murray Blum told the press that *Solenopsis invicta* "is totally out of control. The South has been conquered."[231]

Conclusion

The translation of an idea into action is fraught with difficulties, and that is especially true when, as in the cases here, the ideas themselves are only partial representations of a wider, more complex world. In this chapter, I have traced how the USDA and its opponents attempted to make their ideas into public policy. The two sides of the fire ant wars took turns seeing their visions made reality, but even at the height of their policymaking prowess, neither ever fulfilled its promise.

The USDA was one of the strongest federal agencies, and proponents of the eradication program had forged an iron triangle with lobbying groups in the South and Jamie Whitten's committee in Congress that nurtured the imported fire ant program. But, despite the USDA's power, problems with the program accumulated. The PPC entomologists chose to ignore them, worried that any acknowledged deficiencies might be deemed a cause to stop the program. Eventually, though, the problems caught up with the agency, and did almost bring the program to a halt, as it became increasingly obvious that the chemical insecticides killed wildlife and contaminated milk and that support for an eradication program was not universal. Ironically, Clifford Lofgren helped save the program with research that the PPC had deemed to have no practical value.

By the time that all of the necessary conditions for another eradication program had again fallen into place, though, the political landscape

229. S. D. Porter and D. A. Savignano, "Invasion of Polygyne Fire Ants Decimates Native Ants and Disrupts Arthropod Community," *Ecology* 71 (1990): 2095.

230. R. S. Devine, *Alien Invasion: America's Battle with Non-Native Animals and Plants* (Washington, D.C.: National Geographic Society, 1998), 129.

231. "Red Fire Ant Conquers South, Is Now 'Totally out of Control,'" *Houston Chronicle*, 1 December 1983, 12E.

had changed. Environmentalists created new institutions to check the USDA's power. They banned the insecticides used to eradicate the imported fire ant. But the environmentalists' policies also failed to match their promise. The EPA and the EDF concentrated on banning insecticides, protecting wildlife and human health. They succeeded to some extent, saving much of the country from exposure to dangerous chemicals. The environmentalists' limited focus, however, blocked the development of alternative methods of insect control. And their policies misjudged the danger posed by the fire ant. As the polygynous colonies became increasingly dominant, the ant loomed as an ever-larger threat to the South's biodiversity.

Chapter Five: The Practice of Nature, 1978–2000

A man is a bundle of relations, a knot of roots, whose flower and fruitage is the world.

Ralph Waldo Emerson, "History"

Nature is a trinity. It is a thing that exists independently of humans, structuring the world outside of us, structuring the world inside of us, and connecting both. It is also in our heads. "Ideas of nature," writes the cultural historian Raymond Williams, "are the projected ideas of men" (and, I would add, women).[1] Finally, nature is a practice. It is something that we do.[2] We help to create the world around us, even if it always remains beyond our control and subject to laws that we cannot change. These three natures are not distinct from one another. They interrelate, and although it is easier to think—and to write—about them separately, we live them all at once. Nature inspires thoughts; humans see the world and themselves through lenses shaped by their place in history and society; we try to change the world, but our ideas are only imperfectly translated into action and applied to a world that is changing and always beyond our full kenning.[3]

In chapter 4, I explored how the various sides in the fire ant wars endeavored to make their ideas into public policy. In this chapter, I examine the consequences of those actions, the results of bringing together nature, ideas about nature, and ways of interacting with nature. What worked? What failed? And what were the repercussions for humans and the ant—for society and nature?

By the late 1970s, the eradication program of the U.S. Department of Agriculture (USDA) had finally collapsed, and while the Environmental Protection Agency (EPA) succeeded in banning dangerous chemicals, its program also failed to realize its full promise. These policy failures, I suggest, resulted from the friction between nature, ideas about it, and

---

1. R. Williams, *Problems in Materialism and Culture* (London: Verso, 1980), 82.

2. On this point, see R. White, *The Organic Machine: The Remaking of the Columbia River* (New York: Hill and Wang, 1995); R. White, "'Are You an Environmentalist or Do You Work for a Living?': Work and Nature," in *Uncommon Ground: Rethinking the Human Place in Nature,* ed. W. Cronon (New York: Norton, 1996), 171–85.

3. On this point, see A. F. McEvoy, "Toward an Interactive Theory of Nature and Culture: Ecology, Production, and Cognition in the California Fishing Industry," *Environmental Review* 11 (1987): 289–305.

practices of nature. The imported fire ant continued to expand its range, and the environmental movement continued to gain power, the two combining to eventually end the careers of the eradication ideal and the Plant Pest Control Division. But, the ant never became the well-behaved citizen that the environmentalists had imagined it would be, and the EPA's power waned, part of a backlash against the growing environmental movement and a more general decline of confidence in government.

Out of the wreckage of these programs emerged new ideas about the ant and new ways of interacting with it. The image of the insect regained some of its previous ambiguity and entomologists regained tools for its control that often had been ignored in years past. Both the ideas about the ant and the new methods for controlling it hold promise. We may even someday eradicate the ant. But whether *Solenopsis invicta* is eradicated or not, the history of the fire ant wars should be a lesson in humility. We should now know to expect the unexpected, for our ideas, however powerful, are usually incomplete; their translation to practice is difficult; and the nature with which we are working—like ourselves and our society—is always changing and often impossibly complex. As the poet W. H. Auden wrote, "We can only do what it seems to us we were made for, look at this world with a happy eye but from a sober perspective."[4]

Red Ants and the Red Queen

In 1957, when the USDA first announced its intention to eradicate the imported fire ant, the insect occupied about 20 million acres. In the late 1960s, when the department considered a second eradication program, the ant's range covered about 100 million acres. By the end of that eradication program, the imported fire ant had spread to cover almost 200 million acres.[5] Over the twenty-five years between 1957 and 1982, the USDA had spent about $108 million in the attempt at controlling the ant, while state and local governments contributed about another $67 million.[6] By any measure, the eradication programs proved dismal failures. Why?

The principal reason for the overall programs' failure is that the eradication plan of the Plant Pest Control Division (PPC) relied on a

---

4. W. H. Auden, *City without Walls and Other Poems* (New York: Random House, 1969), 28.

5. For an overview of the ant's spread, see A. A. Callcott and H. L. Collins, "Invasion and Range Expansion of Imported Fire Ants (Hymenoptera: Formicidae) in North America from 1918–1995," *Florida Entomologist* 79 (1996): 238–51.

6. M. K. Hinkle, "Impact of the Imported Fire Ant Control Programs on Wildlife and the Quality of the Environment," in *Proceedings of the Symposium on the Imported Fire Ant,* ed. S. L. Battenfield (Washington, D.C.: Government Printing Office, 1982), 140.

simplified vision of both society and nature. All land was considered to be the same: it had to be sprayed with insecticides, regardless of use or ownership. All people had to join the campaign against the ant: the program had to be mandatory, not voluntary. Moreover, the PPC plan rested on the conviction that the ant's essential quality—besides being a pest—was its susceptibility to insecticides: research on its biology, natural history, and habits was of limited practical value. The vision was elegant and powerful. It showed exactly what needed to be accomplished to exterminate the ant and it served as a beacon, attracting support from both the public and politicians. But it disguised a great deal of complexity. Both nature and society are more heterogeneous, more unpredictable, more complicated than the USDA's plans could account for. The PPC's vision outstripped reality, and for all their political clout, agricultural officials could not remake both society and nature enough to match their ideas.[7] Their resources, though impressive, were inadequate for the task; all land could not be treated the same; and the ant proved more resilient than the entomologists expected. The USDA's program, I contend, withered from the heat generated by the friction between the natural world, ideas about nature, and human practices.

The department's plan to eradicate the ant seemed sound. Kill the insect on the periphery of its range and establish a quarantine to prevent its spread. Then, with the insect surrounded, spray the more heavily infested areas until no imported fire ant existed in the United States. The USDA, however, for all it bureaucratic strength, lacked the money and manpower to succeed. During the first eradication campaign, funding levels limited the entomologists to treating 1 million acres per year—and no more than two hundred thousand acres in any one state.[8] At that rate, it would have taken fifty years to exterminate the pest from Alabama alone, and even longer when one factored in the difficulties people had transforming a dream into reality.[9] "We recognized from the beginning that, regardless of treatment, we must expect some survival because, under field conditions, we, as individuals or collectively, cannot obtain complete perfection, and the machines used for

7. On the difficulties of applying all-encompassing ideas to a complex world, see J. C. Scott, *Seeing Like a State: How Certain Schemes to Improve the Human Condition Have Failed* (New Haven: Yale University Press, 1998).

8. L. J. Padget, "Some Facts about the Imported Fire Ant Program and Reported Wildlife Losses," 20 June 1958, fire ant file, box 198, Plant Pest Control Division papers, RG 463, National Archives and Records Administration, College Park, Md. (hereafter PPC papers).

9. On this point, see Claude Kelly's remarks on N. Gannon and G. C. Decker, "Insecticide Residues as Hazards to Warm-Blooded Animals," *Transactions of the Twenty-fourth North American Wildlife Conference*, 1959, 131.

applying granules are far from being everything we desire," C. C. Fancher, head of the PPC's southern regional office, told E. D. Burgess, head of the PPC, in 1960.[10] Fancher estimated that up to one-third of the treated areas would need to be sprayed again, effectively raising the area that needed treatment from 20 million to 26 million acres.[11] At that rate, many areas would go decades without the protection afforded by insecticides, giving the ant places to reinfest. Like the Red Queen in *Through the Looking Glass,* who always ran to stay in the same place, the PPC would always be spraying to eradicate the ant. "We should have realized . . . that the rate of treatment in the various states is not sufficient at the present time to clear the area in front of previously treated blocks rapidly enough to prevent them presenting hazards of reinvasion to blocks treated already," Fancher acknowledged in 1961.[12]

Inadequate resources, though, could be overcome—and Congress did its best to help the USDA, increasing the program's funding as the century wore on. More damning for the eradication program was the PPC entomologists' simplified view of the southern landscape. The entomologists, I have no doubt, knew that the South was a complex place, a mosaic of landscapes inhabited by a pluralistic people. The eradication program, however, could not take account of this diversity. All land had to be sprayed, regardless of use or ownership. In practice, however, this ideal could not be achieved. To protect wildlife, the USDA had to skip swamps and woodlands. The fire ant avoided these areas, too, but not completely. Harlow Mills noted that the ant occasionally inhabited swamps and woodlands, and Fancher admitted that skipping these places made eradication nearly impossible to achieve.[13] Only a few ants lived in undisturbed wilderness, but it only took a few to repopulate the South and mock the PPC's eradication ideal. Ownership of the land also mattered, for despite the USDA's publicity campaign, some people refused to have their land sprayed.[14] Do not

10. C. C. Fancher to E. D. Burgess, 19 September 1960, fire ant files (1960), PPC papers.

11. Ibid.

12. C. C. Fancher to E. D. Burgess, 3 January 1961, fire ant files (1961), PPC papers.

13. On the ant's presence in these places, see H. B. Mills to W. L. Popham, 29 December 1958, man and animals imported fire ant file, box 1, Entomology Research Division: Entomology Director's Correspondence, 1959–1965, 1055 (A1), Agricultural Research Service papers, RG 310, National Archives and Records Administration, College Park, Md. (hereafter ARS papers). For Fancher's admission, see C. C. Fancher to E. D. Burgess, 3 January 1961, fire ant files (1961), PPC papers.

14. For examples, see the five boxes of letters in "Plant Protection Program: A Sample of Incoming Correspondence from the Public Relating to the Use of Mirex in the Imported Fire Ant Program, Received in the Fall and Spring of Fiscal Year 1971," ARS papers. Most correspondents opposed the program, although not all objectors came from the South.

presume that "the property owner is willing to cooperate in the very doubtful benefits of this program," one landowner told the USDA. "I consider this as an unwarrantable invasion of my property rights."[15] The ant could use these untreated areas as a sanctuary from the poisons used to kill it. Southern society and southern nature proved less malleable than the PPC's plans allowed and this masked complexity helped bring the program to ruin. Even with unlimited resources, it is impossible to imagine how the USDA could have eradicated the ant while not spraying these areas.

Finally, I think that the eradication program failed because the PPC entomologists did not reckon with the biology of the imported fire ant. Evolved on a South American flood plain, the ant thrived in areas of disturbance, just like those created by the department's campaign. The insecticides worked like the Río Paraguay and the Río Paraña in South America, or like the bulldozer revolution in the southern United States, creating the very conditions favored by the imported fire ant. Thus, not only did the program fail, leaving sanctuaries for the insect, it helped to spread the ant (fig. 16). No studies prove the case, but the evidence is suggestive, especially for the Mirex program. Combining field observations with a simple mathematical model, William Buren and some colleagues showed that a bait treatment that killed 95 percent of an ant population—close to the efficacy of Mirex—allowed the disturbance-loving *Solenopsis invicta* to replace other ants. An ant community composed of 1 percent *Solenopsis invicta* and 99 percent other species inverted itself. Less than three years after application of the bait, imported fire ants made up 90 percent of the community. If, on the contrary, no Mirex were used, it would take the ant thirty years to achieve that level of dominance.[16]

Clifford Lofgren's later research supported Buren's speculations. In one study he found that forty-four weeks after he applied bait to a field, the number of imported fire ant mounds skyrocketed by 327 percent.[17] The insect overwhelmed the disrupted areas, new nests proliferating as they would after a river's flood or in the shadow of the bulldozer

15. V. A. Norman to USDA, 29 April 1959, complaints-2-fire ant file, box 211, PPC papers.

16. W. F. Buren, J. L. Stimac, and F. G. Maxwell, "A Model of Fire Ant Populations: Implications toward a Prolonged Problem," copy in Murray S. Blum papers, in author's possession (hereafter Blum papers).

17. C. S. Lofgren and D. F. Williams, "Red Imported Fire Ants (Hymenoptera: Formicidae): Population Dynamics following Treatment with Insecticidal Baits," *Journal of Economic Entomology* 78 (1985): 863–67. See also C. T. Adams, T. E. Summers, C. S. Lofgren, D. A. Focks, and J. C. Prewitt, "Interrelationship of Ants and the Sugarcane Borer in Florida Sugarcane Fields," *Environmental Entomology* 10 (1981): 417.

Figure 16. These ants embodied the name that Buren gave them. They were unconquerable, bloody but unbowed, eating Mirex, but not dying. This editorial cartoon ran in *The Reflector*, Mississippi State University's paper, which also published Denzel Ferguson's first attack on the Mirex program. The caption reads, "Mirex and corn cobs, Mirex and corn cobs, for two years–nothing but Mirex and corn cobs." (From general info to 1970 file, Mississippi Department of Agriculture and Commerce papers, Mississippi Department of Archives and History.)

revolution. Lofgren pointed out that the actual number of ants had decreased, suggesting that some control had been achieved, but he could not ignore the huge leap in the number of colonies. The smaller number of ants was to be expected, anyway. The colonies were young and so had relatively few members. "The implications are clear," Walter Tschinkel said. "Large-scale, unspecific control programs . . . actually aid rather than hinder the establishment and spread of the fire ant and accentuate its dominance over native ants."[18] The eradication advocates' dilemma was worse than the Red Queen's. Running as fast as they could, the federal entomologists still lost ground.

The PPC's plan for eradication obscured the ant, the land, and southerners behind abstractions. Reality was more complicated than the entomologists allowed, and when their ideas, imperfectly translated into action, encountered nature in all its complexity, new and

18. W. R. Tschinkel, "The Ecological Nature of the Fire Ant: Some Aspects of Colony Function and Some Unanswered Questions," in *Fire Ants and Leaf-Cutting Ants: Biology and Management*, ed. C. S. Lofgren and R. K. Vander Meer (Boulder. Westview, 1986), 75.

unexpected consequences resulted: the red imported fire ant continued, maybe even accelerated, its spread. This continued spread—whether facilitated by the insecticides or the result of the USDA's inability to see its ideas fully realized—in turn contributed to the declining fortunes of the PPC and the eradication ideal. The ant did not—could not—press the case itself. Yet it came to represent another of the PPC's failure. In 1978, the *Washington Post* reported, "Not since the Dust Bowl days of the 1930s has the United States been so assaulted by pests. The Gypsy moth in the Northeast, the fire ant in the Southeast, the corn borer in the Midwest and the grasshopper in the Plains states now infest the nation in record numbers."[19] Agricultural officials argued that the banning of many insecticides during the 1970s had caused the irruptions, but that claim missed the point.[20] Outlawing the chemicals may have contributed to the outbreak, but the insects had resisted eradication for decades, even when no laws barred the USDA from using its weapons of choice. Combined with the rise of the environmental movement, these failures proved the inadequacy of the eradication ideal and helped force a change in the USDA's entomological programs.

In the late 1960s, as the environmental movement was gaining sufficient power to challenge the PPC, entomologists associated with state universities articulated a different vision of entomology. These researchers suggested that their science should embrace what they called "integrated pest management," using not only chemical insecticides, but also biological and cultural control—not unlike the entomological practices that had been abandoned in the wake of DDT and related chemicals.[21] These entomologists were not so much reinventing their science as recovering lost tools. Leo Newsom was one of the idea's leading proponents.[22] William Buren, too, advocated integrated pest management, arguing, "The lessons of the past seem clear that single-minded chemical strategies against the imported fire ants have not worked and will not work. What is difficult to mentally grasp is that it is quite possible to win most or all of the numerous small battles and still lose a war through deficiencies in the overall grand strategy. Nature's own biological factors and constraints

19. T. O'Toole, "Insect Pests Making Global Comeback," *Washington Post,* 6 August 1978, A1.

20. On the USDA's explanations and support for such claims, see ibid.; "The Year of Insects," *Wall Street Journal,* 15 August 1978, 20.

21. For an overview of integrated pest management, see M. Kogan, "Integrated Pest Management: Historical Perspectives and Contemporary Developments," *Annual Review of Entomology* 43 (1998): 243–70.

22. See L. D. Newsom, "The Next Rung up the Integrated Pest Management Ladder," *Bulletin of the Entomological Society of America* 26 (1980): 369–74.

are far more effective in maintaining fire ant populations at a low level than anything devised by man."[23] Eventually, the supporters of integrated pest management gained the favor of the Nixon administration and forced changes in the USDA's entomological programs. By 1979, historian Paolo Palladino writes, the "proponents of 'integrated control' appear to have won the day; they had marshaled a far more powerful coalition than the Department of Agriculture ever could, and were then able to force the Department to adopt their own agenda for pest control."[24]

These changes influenced the operations of the imported fire ant program. When Nixon's environmental advisors reviewed the imported fire ant program after the Environmental Defense Fund (EDF) brought suit in 1970 against the USDA, they decided that the department's inability to control the ant stemmed from its refusal to invest in research on the insect and so recommended that the department spend more money on biological research and less on control.[25] In 1971, the USDA, probably responding to this pressure, dispersed $1 million to universities for basic research on the ant.[26] Over the next decades methods to control the insect expanded; what had been discouraged—studying biological and cultural control in addition to chemical control—became standard operating procedure. In 1981, for example, Clifford Lofgren, William Banks, and another USDA entomologist wrote, the imported fire ant "presents a classic imported pest problem in which the pest has been freed of natural enemies. The importation and establishment of a complex of natural enemies needs to be explored as a means of ameliorating the problem."[27] Insect-growth regulators reached the market, although they never revolutionized control of the ant.[28] At the end

23. Quoted in R. Wolf, "Fire Ants," *Florida Naturalist,* December 1978, 11.

24. P. Palladino, "On 'Environmentalism': The Origins of Debates over Policy for Pest-Control Research in America, 1960–1975," in *Science and Nature: Essays in the History of the Environmental Sciences,* ed. M. Shortland (Oxford: British Society for the History of Science, 1993), 209.

25. On the Nixon administration's consideration of increased research, see D. B. Rice to J. C. Whitaker, 8 March 1971, pesticides—1971 file [2 of 3], box 91, John Whitaker subject files, White House Central Files, Richard M. Nixon papers, National Archives and Records Administration, College Park, Md.

26. On the increased funding, see C. S. Lofgren, "History of Imported Fire Ants in the United States," in *Fire Ants and Leaf-Cutting Ants: Biology and Management,* ed. C. S. Lofgren and R. K. Vander Meer (Boulder: Westview, 1986), 43; M. S. Blum to W. S. Creighton, 21 September 1971, William Steel Creighton papers, an unprocessed box of material incorporated in the uncatalogued E. O. Wilson papers, Library of Congress, Washington, D.C. (hereafter Creighton papers).

27. D. P. Jouvenaz, C. S. Lofgren, and W. A. Banks, "Biological Control of Imported Fire Ants: A Review of Current Knowledge," *Bulletin of the Entomological Society of America* 27 (1981): 207.

28. D. F. Williams, H. L. Collins, and D. Oi, "The Red Imported Fire Ant (Hymenoptera: Formicidae): An Historical perspective of Treatment Programs and the Development of Chemical Baits for Control," *American Entomologist* 47 (2001): 153.

of the twentieth century, the Agricultural Research Service laboratory where studies on the ant were carried out noted, "The Center conducts research on insects of agricultural, medical, and veterinary importance with the goal of achieving control of pest species through the development of environmentally acceptable approaches. Emphasis is placed on developing components and systems for integrated pest management, based upon an understanding of the behavior, physiology and ecology of pest species."[29]

An indication of how far the eradication ideal had fallen could be found in Jackson, Mississippi's *Clarion-Ledger,* which reported in 1996, "'Eradication' is a dirty word among the small corps of fire ant researchers."[30] Elsewhere, one entomologist said, "The worst thing we ever did was use the word 'eradicate.' We couldn't eradicate this thing with a bomb."[31] Another federal entomologist admitted that expectations needed to be diminished. The battle against the ant would be a long one. "There's no silver bullet. No magic wand," he said.[32] Even Jim Buck Ross quit using chemicals to control the ant on his property. "I got 'em knee high on my place. I don't do anything," he said in 1988.[33]

The Ironies of Environmentalism

"Guess what?" the Texas A&M University entomologist F. W. Plapp asked the Environmental Defense Fund in 1976. "We've won one! John White, the (elected) Texas Commissioner of Agriculture felt the heat and announced last week that there will be no more Mirex spraying in Texas for fire ant control. . . . I'd love to take credit for this, but we can't. It was really the Lufkin, Texas environmentalists who did all the hard work and carried the day."[34] Two years later, environmentalists nationwide celebrated when the EPA banned Mirex; and five years later they basked in their long-sought victory against ferriamicide. They had made it illegal to spread the carcinogenic chemicals across the South, protecting wildlife and human health. They had also banned a host of

29. See www.nps.ars.usda.gov/locations/locations.htm?modecode=66-15-10-00 (accessed 14 May 2003).

30. B. Reid, "Fire Ants: Creatures Biggest Pest in South," *Clarion Ledger,* 9 June 1996, 1A.

31. "Fire Ants: The Experts Throw up Their Hands," *U.S. News and World Report,* 17 January 1977, 79.

32. R. L. Brown, "Fire Ants: Buggy Battalions Beating Mere Mortals," *Atlanta Journal-Constitution,* 21 June 1992, M2.

33. A. Huffman, "Environment May Pay the Price in Fire Ant War," *Clarion Ledger,* 4 November 1988, 1A.

34. F. W. Plapp to M. Hinkle, 5 October 1976, correspondence, internal file, box 71, Environmental Defense Fund papers, MS 232, State University of New York, Stony Brook, N.Y.

other chemical insecticides, including chlordane, dieldrin, heptachlor, and DDT. There was much to be proud of.

The victory, though, was alloyed with loss. Environmental policy, like agricultural policy, corresponded only imperfectly with the natural world that it meant to manage. According to the EPA's diagnosis, the threats to the environment stemmed from the indiscriminate use of insecticides. The solution was to keep the chemicals out of the environment so that nature could reassert its balance. But the profligate use of insecticides was actually a symptom of a deeper malady—the same condition, in fact, that allowed the imported fire ant to irrupt and become a pest. The root cause of both problems was the new land-use practices that remade the South in the years after World War II. Modern industrialized farming—because it relies on densely planted monocultures that can fuel insect irruptions—made insecticides a necessity, and the bulldozer revolution created a plethora of habitats in which the ant could outcompete other insects. Banning insecticides helped preserve human health and won the political support of the environmental movement, but it neither ended America's dependence on pesticides nor allowed the fire ant to become a nonthreatening member of a balanced southern ecosystem.

After the conclusion of the Mirex program, the ant kept spreading across the Sun Belt. It moved through the South and west into Texas, New Mexico, Arizona, Nevada, and California, as far north as Sacramento.[35] It sailed to Puerto Rico.[36] It established beachheads in Virginia and Maryland.[37] Exploiting the disrupted environments that increased land development created in abundance—and polygynous colonies building unexpectedly dense communities—the ant continued to pester, annoy, and harass people, eating some crops, occasionally killing wildlife, and stinging many people. In North Carolina, for example, farmers were losing about a bushel of soybeans per acre to the pest.[38] And a 1994 survey of veterinarians found that Texans spent $750,000 treating fire ant injuries to animals.[39] Three decades after the Audubon Society's Harold Peters proclaimed that the southern ecosystem had accommodated the insect, a public health report noted, "Residents in the southeastern United States would hardly describe life

35. S. W. Taber, *Fire Ants* (College Station: Texas A&M University, 2000), 219; "Is It 'Them!'?" *Sacramento Bee*, 3 May 2000, A4.

36. W. F. Buren, "Red Imported Fire Ant Now in Puerto Rico," *Florida Entomologist* 65 (1982): 188–89.

37. S. W. Taber, *Fire Ants* (College Station: Texas A&M University, 2000), 219.

38. Ibid., 195.

39. J. Grisham, "Attack of the Fire Ant," *Bioscience* 44 (1994): 589.

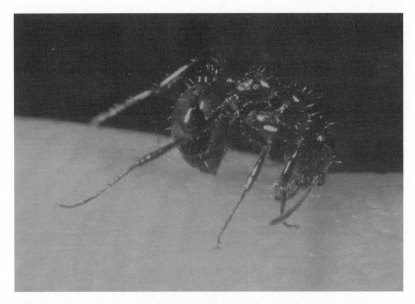

Figure 17. Photograph of a fire ant crawling on someone's hand or arm. The imported fire ant is tiny, but its sting makes it a large problem. A small group of people are allergic, and the proteins in its venom can cause an anaphylactic reaction. For others, the sting only results in a welt, but it is still uncomfortable—"like lighting a match too near the skin," in E. O. Wilson's phrase. The problem is exacerbated when an entire mound attacks. In the 1990s, several nursing home residents died after being stung more than a thousand times. (Photo courtesy of Murray S. Blum.)

with the aggressive imported fire ant as peaceful coexistence."[40] At the end of the century, scientist Stephen Taber estimated that about 80,000 people sought medical attention for ant stings each year (fig. 17).[41]

No longer overshadowed by the dangers of heptachlor or Mirex, the ant's negative characteristics came into sharp relief, and wildlife enthusiasts remembered their disdain for the insect. The number of hunters across the South had declined, but interest in native animals had not, and people worried that the red imported fire ant destroyed indigenous fauna. "Ask 100 Texas landowners and quail hunters what the birds' worst enemy is (excluding weather)," noted the *Houston Chronicle* in 1995, "and 99 of them will immediately answer 'fire ants!'"[42] Wildlife biologists again crossed swords about the importance of this predation, some claiming the ant reduced animal numbers significantly,

40. S. F. Kemp, R. D. deShazo, J. E. Moffitt, D. F. Williams, and W. A. Buhner II, "Expanding Habitat of the Imported Fire Ant *(Solenopsis invicta):* A Public Health Concern," *Journal of Allergy and Clinical Immunology* 105 (2000): 683.

41. S. W. Taber, *Fire Ants* (College Station: Texas A&M University, 2000), 130.

42. S. Tompkins, "Quail Prospects Look Bleak," *Houston Chronicle,* 22 October 1995, 29.

others insisting that the destruction was negligible.[43] But however troublesome the ant might or might not have been regionwide, it clearly threatened some specific bird and animal populations, and caused serious problems for some localities. For example, biologists found that the insect reduced the nesting success in some swallow colonies by almost one-third.[44] Mice, too, seemed vulnerable to the insect's predation.[45] Even the Audubon Society reconsidered its opinion of the ant. In the late 1980s, the insect forded the Gulf of Mexico and reached the Rollover Pass Islands off the coast of Texas, where National Audubon Society biologists managed a community of seabirds. The ant population grew to a density of 180 mounds per acre and the insect devoured baby birds as they hatched. Frustrated Audubon officials turned to entomologists for help controlling the ant.[46] According to one report, the entomologists suggested an approach that would control the ant only where it caused problems, not eradicate it. An Audubon official replied, "You don't understand. We want the ants off the island."[47]

As the ant spread across the country and produced yelps of outrage, Americans continued to dump insecticides into the environment: lawns and modern farms required their continued use. In 1980, 1.5 billion pounds of pesticides were used in the United States, twice as much as in 1962, when Rachel Carson published *Silent Spring*.[48] Two decades later, that number remained about the same.[49] And although the EPA had banned the chemicals used in the USDA eradication campaigns, those affected by the ant continued to use other insecticides. Between 1958 and 1998, USDA scientists tested over 7,200 baits. Nine became commercially available, with two—including Mirex—later pulled from the market.[50] In

43. For the debate, see L. A. Brennan, "Fire Ants and Northern Bobwhites: A Real Problem or a Red Herring," *Wildlife Society Bulletin* 21 (1993): 351–54; C. R. Allen, R. S. Lutz, and S. Demarais, "Ecological Effects of the Invasive Nonindigenous Ant, *Solenopsis invicta*, on Native Vertebrates: The Wheels on the Bus," *Transactions of the 63rd North American Wildlife and Natural Resources Conference*, 1998: 56–65.

44. M. J. Killion and S. B. Vinson, "Ants with Attitudes," *Wildlife Conservation* 98 (January–February 1995), 48.

45. Ibid., 50–51.

46. "Fire Ant Attacks Maim South Texas Wildlife," *Dallas Morning News*, 8 July 1989, 24A.

47. R. Conniff, "You Never Know What the Fire Ant Is Going to Do Next," *Smithsonian*, July 1990, 57.

48. C. J. Bosso, *Pesticides and Politics: The Life Cycle of a Public Issue* (Pittsburgh: University of Pittsburgh Press, 1987), 237.

49. Committee on the Future Role of Pesticides in U.S. Agriculture, *The Future Role of Pesticides in U.S. Agriculture* (Washington, D.C., National Academy Press, 2000), 33.

50. D. F. Williams, H. L. Collins, and D. Oi, "The Red Imported Fire Ant (Hymenoptera: Formicidae): An Historical Perspective of Treatment Programs and the Development of Chemical Baits for Control," *American Entomologist* 47 (2001): 152.

the mid-1990s, the EPA allowed the use of more than 150 chemicals to control the ant.[51] "We spend more on concoctions that are supposed to kill fire ants than we do on groceries," quipped a Georgia newspaper columnist.[52] These chemicals were less toxic and less persistent than their outlawed predecessors, but they too spread through the environment, the repercussions of the percolation unclear.[53] "Right now tens of thousands of people are pouring pesticides out there" to kill *Solenopsis invicta,* said Felder Rushing, a Mississippi extension service horticulturalist, in November 1988. "With everybody poisoning their lawns in Jackson we're talking about big pesticide runoff into the Pearl River."[54]

Southerners also turned to other chemicals not approved by the EPA, anything—even pure bunk—to stop an insect that had not been accommodated as part of the local ecosystem but remained a pest. Some poured gasoline and molten lead into ant nests, a tremendous environmental hazard if used to control the insect in large areas, not to mention expensive.[55] Others sprinkled grapefruit peels or grits around the nests on the faulty assumptions that citrus repelled the ant or grits would swell inside in *Solenopsis invicta*'s stomach and burst the insect.[56] In 1979, Dan Wheeler of Orlando, Florida, told *The Sentinel Star,* "I pour 85 proof bourbon into the ant hills. This makes them drunk, and when they come out for fresh air I have an opportunity to hit them in the head with a hammer."[57]

The EPA, then, had banned the insecticides widely perceived to control the ant, blocked the introduction of alternative control methods, and misjudged the beneficence of the little red creature—leaving it to southerners to control the bug with grits and bourbon. Resentment toward the agency, and the environmental movement generally, grew. Why was the fire ant "taking over the country," asked *Atlanta Journal-Constitution*

51. S. Norvell, "Force of Nature," *Washington Post,* 23 June 1996, W14.

52. J. Minter, "Life in the Exurbs Can Be Blissful—or a Picnic Shared with Fire Ants," *Atlanta Journal-Constitution,* 6 October 1994, M2.

53. Committee on the Future Role of Pesticides in U.S. Agriculture, *The Future Role of Pesticides in U.S. Agriculture* (Washington, D.C., National Academy Press, 2000), 3.

54. A. Huffman, "Environment May Pay the Price in Fire Ant War," *Clarion Ledger,* 4 November 1988, 1A.

55. On gasoline, see W. Minis, "Nearly Invincible Fire Ants Villains of Nature," *Charleston (S.C.) Post and Courier,* 3 June 2002, 1D; on molten lead, see "Fire Ant Extinguisher," *Sarasota (Fla.) Herald-Tribune,* 23 June 1997, 8A.

56. On grapefruit peels, see D. Gill, "Fighting Fire Ants; It's a War That Persistence Can Win," *Times-Picayune,* 15 May 1998, E1; on grits, see E. Yoffe, "The Ants Come Marching in; Large Areas of the South and Southwest Have Fallen to Fire Ant Invaders," *Washington Post,* 23 August 1988, Z8.

57. " 'Better Way' Suggested to Get Rid of Fire Ants," 30 June 1979, Blum papers.

columnist Jim Minter. "The usual," he answered. "Environmentalists decided we'd be better off letting the ants eat us than risk killing a few birds or fish."[58] It was now the EPA, not the USDA, that figured as the autocratic bureaucracy unconcerned with individual liberties. "Meet the new boss," sang the rock-n-roll group The Who in 1971. "Same as the old boss."[59]

Barron's, the business magazine, opined that bureaucratic red tape made it virtually impossible to fight *Solenopsis invicta*, leaving southerners no choice but to suffer the ant; and the *Wall Street Journal* called 1978 the "year of the insect," editorializing that the outlawing of Mirex, coupled with the banning of other insecticides, "may well mark the first major success of the radical fringe of the environmental movement in giving the country back to the bugs. Fire ants, which destroy cropland with their nests and attack livestock and even people, are rampant in Mississippi; EPA rules and a lawsuit by the Environmental Defense Fund have blocked the state from controlling the ants."[60] Carl Djerassi, president of the biotechnology company Zoecon, had made a similar point several years earlier. The USSR—the "evil empire"—had talked to his company about purchasing insect-growth regulators, he said. Even a communist country recognized that the free market was the most efficient producer of the technology, but the EPA was blind to the fact, a bureaucratic impediment.[61] John Quarles, the EPA deputy administrator, conceded the problem. "I am a federal bureaucrat," he began his 1976 book *Cleaning up America*, sarcasm masking a real problem. "I am one of those bureaucrats everyone complains of, the ones who frustrate worthy government programs and tangle them in endless red tape. I am part of the faceless, gray government machine."[62]

Politicians impatient over the lack of weapons against the spreading ant tried to dismantle the agency and unravel the skein of laws that they perceived as a restriction on American freedoms. In 1978, Tom DeLay, an exterminator from Houston, joined the Texas legislature, frustrated by the

58. J. Minter, "Send Them Back to Alabama: Marauding Fire Ants Are Still on the March, Lighting a Blaze under Would-be Legislator," *Atlanta Journal-Constitution*, 26 March 1996, A9.

59. The Who, "Won't Get Fooled Again," *Who's Next* (MCA, [1971] 1995).

60. S. Schiebla, "Fighting the Fire Ant: Regulations Make It Almost Impossible to Do," *Barron's*, 10 May 1976, 11, 21–22; "The Year of Insects," *Wall Street Journal*, 15 August 1978, 20.

61. C. Djerassi to R. Long, 14 February 1974, insects, April 19 to July 16 file, box 5855, General Correspondence, 1906–1975, PI 191 and 1001 A-E (UD), Secretary of Agriculture papers, RG 16, National Archives and Records Administration, College Park, Md.

62. J. Quarles, *Cleaning up America: An Insider's View of the Environmental Protection Agency* (Boston: Houghton Mifflin, 1976), xi.

attack on Mirex.[63] He compared the EPA to the Gestapo.[64] The following year, Georgia representative Dawson Mathis—who later lobbied against ferriamicide for American Cyanamid—introduced legislation that reauthorized the use of Mirex and gave Congress veto power over any EPA pesticide regulation—until 1985, when the agency would lose all authority to regulate the chemicals. Mathis's colleague, Billy Lee Evans said, "Since the cessation of the use of Mirex, almost daily people are telling me the fire ants are back. This is a very big issue with my people. We are at a point where we have to have some assistance."[65] The American Farm Bureau championed Mathis's bill because "the EPA has been derelict in its response to petitions for relief from the fire ant problem."[66]

Antagonism toward the EPA was in part a response to the continued spread of *Solenopsis invicta*, but it was also a side effect of the way that federal bureaucracies had grown in the years after World War II and the increasingly close relationship between science and politics. By the late 1970s, animosity toward the environmental movement and the EPA's antagonistic relationship with industry galvanized a countermovement.[67] The House of Representatives voted down Mathis's proposal to reauthorize the use of Mirex, but "by a somewhat impressive 288-121 margin"—in the words of political scientist Christopher Bosso—the House approved the veto provision, evidence of growing regret about the power ceded to the EPA.[68] In 1980, Ronald Reagan swept Jimmy Carter from office and brought a determined disregard for environmental laws to the White House. Tom DeLay rode Reagan's coattails into the U.S. Congress.[69] New EPA administrators promised a less adversarial relationship with industry, a change that environmentalists equated with surrendering many of the gains they had made over the past decades.[70] To alter the relationship between the agency and industry, though,

63. M. Weisskopf and D. Maraniss, "Forging an Alliance for Deregulation; Rep. DeLay Makes Companies Full Partners in the Movement," *Washington Post,* 12 March 1995, A1.

64. P. Perl, "Absolute Truth; Tom DeLay is Certain That Christian Family Values Will Solve America's Problems; But He's Uncertain How to Face His Own Family," *Washington Post,* 13 May 2001, W12.

65. W. Sinclair, "Battle against Fire Ants Heats up over Pesticides," *Washington Post,* 13 October 1979, A2.

66. V. R. Glasson to House Members, 9 October 1979, Blum papers.

67. On the backlash, see S. P. Hays, *Beauty, Health, and Permanence: Environmental Politics in the United States, 1955–1985* (New York: Cambridge University Press, 1987), 491–526.

68. C. J. Bosso, *Pesticides and Politics: The Life Cycle of a Public Issue* (Pittsburgh: University of Pittsburgh Press, 1987), 205.

69. M. Weisskopf and D. Maraniss, "Forging an Alliance for Deregulation; Rep. DeLay Makes Companies Full Partners in the Movement," *Washington Post,* 12 March 1995, A1.

70. C. J. Bosso, *Pesticides and Politics: The Life Cycle of a Public Issue* (Pittsburgh: University of Pittsburgh Press, 1987), 210–11.

required increasing government power, not diminishing it, since environmental sentiment remained strong. "If the [Reagan] administration were to forge rapid and fundamental change," historian Samuel Hays explains, "it would do so not through Congress or the courts but through executive action. New policymakers would bring about new policy and enhance executive authority in the face of the courts, Congress, and the agencies."[71] The administration marshaled its own experts to counter those associated with the environmental movement.

This continual expansion of the federal government, when coupled with the increasing alliances between the bureaucracies and scientists, historian Brian Balogh shows, had a corrosive effect on public confidence in government experts, including those with the EPA. As more experts entered an increasingly competitive policy arena, Balogh argues, they turned to the public to support their agendas. But making their case publicly had an unintended consequence. "As each new participant entered the fray equipped with its own 'independent' experts, the once exalted political clout of expertise waned."[72] The public came to see experts not so much as defenders of truth as another species of politician. Trust in authority declined. Already by 1969, the political scientist Theodor Lowi had detected a "crisis in public authority."[73] By taking an adversarial relation toward agriculture—and industry—the EPA had won supporters and banned toxic chemicals, but it also ruined its own credibility. In 1990 three political scientists noted, the EPA "communicated the message that each interest or agency can appropriately pursue its own goals with no broader view of the national interest."[74] The continued spread of the ant proved to many that the EPA was concerned only about its own bureaucratic status; it cared nothing for the public. "Fire ants," DeLay sneered in 2001. "You can thank the EPA for those!"[75]

Red Ants in Gray Hats

After the eradication ideal crashed and as environmentalists battled those who wanted to reverse their gains, a new vision of the imported

71. S. P. Hays, *Beauty, Health, and Permanence: Environmental Politics in the United States, 1955–1985* (New York: Cambridge University Press, 1987), 493.

72. B. Balogh, *Chain Reaction: Expert Debate and Public Participation in American Commercial Nuclear Power, 1945–1975* (Cambridge: Cambridge University Press, 1991), 303.

73. T. J. Lowi, *The End of Liberalism: Ideology, Policy, and the Crisis of Public Authority* (New York: Norton, 1969).

74. M. Landy, M. Roberts, and S. Thomas, *The Environmental Protection Agency: Asking the Wrong Questions* (New York: Oxford University Press, 1990), 280.

75. P. Perl, "Absolute Truth; Tom DeLay is Certain That Christian Family Values Will Solve America's Problems; But He's Uncertain How to Face His Own Family," *Washington Post,* 13 May 2001, W12.

fire ant emerged—and a new way to control it. In the 1950s, the USDA had stripped the insect of any ambiguity, making the ant a pest. In 1962, Carson reversed the polarity of opinion about the ant, downplaying its negative aspects to make it an important part of American nature. Both of these images retained credibility during the remaining years of the century. In 1988, for example, *Texas Monthly* named the ant "public enemy number one."[76] Nine years later, entomologist S. Bradley Vinson said that this image of the imported fire ant "probably stated best what most people in the infested region consider this creature."[77] That same year, on the other hand, author Brenda Raudenbush sanctified the insect in her book *Brilly and the Boot*, using the ant to explore Christian virtues.[78] A more nuanced view of *Solenopsis invicta* also took shape, however, as southerners accepted that they would have to live with the insect. It had become part of the natural world, but a much-loathed part. The entomologist Winfield L. Sterling captured the emerging sentiment in 1978 when he said that the ant was neither a villain nor a hero. It wore neither a black hat nor a white one. The red imported fire ant wore a gray hat.[79] It ate crops, killed wildlife, stung, and altered the ecology of the region that it invaded. But it also ate boll weevils—and Sterling even suggested that the insect be used to control the cotton pest.[80]

This conflicting image of the ant—as both good and evil—gained currency in popular culture. In the 1976 novel *The Fire Ants*, author Saul Wernick told of a Southern valley where people "can't go out into their fields or backyards or go visit their dead relatives in the cemetery and . . . kids can't play in the school yards" for fear of being stung to death.[81] The ant killed young lovers frolicking in pastures and resisted the might of the U.S. government. The National Guard napalmed the valley, then spread Mirex. But a mutant queen escaped and gave birth to offspring with a sting "strong enough to kill a human being with a single jab."[82] The novel reflected E. O. Wilson's contention that the campaign against the fire ant was the Vietnam of entomology. And yet, the insect was not purely evil. It ultimately redeemed the val-

---

76. E. Yoffe, "The Ants from Hell," *Texas Monthly* 16 (1988): 80–84, 142–43.

77. S. B. Vinson, "Invasion of the Red Imported Fire Ant (Hymenoptera: Formicidae): Spread, Biology, and Impact," *Bulletin of the Entomological Society of America* 43 (1997): 24.

78. B. Raudenbush, *Brilly and the Boot* (Conyers, Ga: Panola Publisher, 1997).

79. W. L. Sterling, "Imported Fire Ant . . . May Wear a Grey Hat," *Texas Agricultural Progress* 24 (1978): 19–20.

80. R. Wolf, "Fire Ants," *Florida Naturalist*, December 1978, 11–12.

81. S. Wernick, *The Fire Ants* (New York: Award Books, 1976), 16.

82. Ibid., 220.

ley, devouring the sadistic henchman who enforced the policies of a morally bankrupt senator. The insect, dangerous though it was, deserved to be cheered—as the University of South Carolina football team demonstrated in 1984, when it nicknamed its defensive unit "The Fire Ants" because of the team's red and black colors and the players' stinging tackles.[83] Pride gilded this name, praise for the ant's resiliency, strength, and bloody, unbowed head. But beneath was a layer of apprehension: the insect stung. The insect was neither a monster nor more beneficial than harmful. It was, as the *Washington Post* said in 1996, "a force of nature" both good and bad, and, finally, beyond human control.[84] The Texas environmentalist Bill Oliver sang, "And they call her Queen Invicta/Fire ant invincible/Nothing could stop her/Predator or chemical."[85]

In the twentieth century, history-minded southerners started to re-enact Civil War battles as a way of coming to terms with their defeat in that war and transform shame into pride.[86] In 1983, the year of ferriamicide's demise, residents of Marshal, Texas, held their first festival in honor of the fire ant, a ritual that I suspect did some of the same cultural work as Civil War reenactments, allowing southerners to confront what journalist Charles Haddad called "The South's other 'Lost Cause,'" and letting participants symbolically gain mastery over the insect.[87] Political leaders in attendance at the festival promised the crowds that one day they would control the insect; but until that time frustrations were vented by the chance festival-goers had to tear apart an oversized model of the ant. And contestants in the chili cook-off laced each pot with at least one *Solenopsis invicta*. Yet, again, mixed with these symbolic triumphs were expressions of reverence: the insect was a worthy foe. The winner of the beauty contest, for example, was crowned Miss Fire Ant.[88] New mythologies made sense of new realities.

83. T. Price, *The '84 Gamecocks: Fire Ants and Black Magic* (Columbia: University of South Carolina Press, 1985).

84. S. Norvell, "Force of Nature," *The Washington Post,* 23 June 1996, W14.

85. Bill Oliver and the Otter Space Band, "Queen Invicta (Fire Ant Invincible)," *Have to Have a Habitat* (Texas Deck Music, BMI, 1995).

86. R. Allred, "Catharsis, Revision, and Re-Enactment: Negotiating the Meaning of the American Civil War," *Journal of American Culture* 19 (1996): 1–13.

87. Quotation from C. Haddad, "Fire Ants Defy Eradication, Burrow into Southern Lore," *The Atlanta Journal-Constitution,* 11 November 1990, B1.

88. For descriptions of the fire ant festival, see E. Yoffe, "The Ants Come Marching In; Large Areas of the South and Southwest Have Fallen to Fire Ant Invaders," *Washington Post,* 23 August 1988, Z8; S. Gamboa, "Fire Ant Festival Offers a Chance for Revenge," *Dallas Morning News,* 3 October 1988, 17A; R. Conniff, "You Never Know What the Fire Ant Is Going to Do Next," *Smithsonian,* July 1990, 48–49.

Reflecting on the controversies over the meaning of the imported fire ant in 1968, William Creighton had pined for a time "when *Solenopsis* can be regarded as just another genus of ants—not a springboard for personal advancement or a means for extracting large sums of money from the government."[89] A jab at both E. O. Wilson and the federal entomologists, Creighton's lament also contained elements of a long-held dream: that humans could shrug off cultural and scientific biases and confront nature directly. The very statement of his wish, however, reflected its impossibility, for even seeing *Solenopsis* as just another genus was to classify and evaluate—actions that already assume a cultural and taxonomic context. There is no way to approach nature free of cultural presuppositions. In the years after the end of the eradication program, people recovered many of the ant's ambiguous characteristics—moving closer, I argue, to a fuller understanding of the ant. But even today our view of the insect remains shaped by our preconceived notions and the position that we occupy within society. As the surprising ascension of the polygynous colonies proves, the ant remains a mysterious and ultimately unpredictable insect. However well we may come to understand the fire ant, the cultural, scientific, and social categories that shape our understanding will never fully contain it.

Decapitating the Ant

Spreading across the South, *Solenopsis invicta* found a new habitat to exploit: entomological laboratories.[90] In years past, studies on insecticides had consumed most of the resources spent on research, but in the wake of the eradication ideal's collapse and the shift toward integrated pest management, entomologists who defined themselves as biologists and wanted to investigate the insect's biology took over. Walter Tschinkel explained, "Seen from the biologist's point of view, the fire ant represents a wonderful research opportunity almost unique in the history of myrmecology. The ant's abundance, ease of maintenance, and general habits make it an outstanding subject for research, while our society's relatively high need for knowledge of this ant gives us the opportunity to carry out this research."[91] The ant entered Tschinkel's laboratory as a tool for exploring the ecology of social insects. William Buren took a job at the University of Florida and studied the ant to find

89. W. S. Creighton to M. S. Blum, 3 April 1968, Creighton papers.

90. On laboratories as habitats, see R. E. Kohler, *Lords of the Fly:* Drosophila *Genetics and the Experimental Life* (Chicago: University of Chicago Press, 1994), 9–11, 19–52.

91. W. R. Tschinkel, "History and Biology of Fire Ants," in *Proceedings of the Symposium on the Imported Fire Ant*, ed. S. L. Battenfield (Washington, D.C.: Government Printing Office, 1982), 35.

biological controls.[92] Murray Blum continued his work on its biochemistry. And at the University of Georgia, Kenneth Ross studied the insect's genetics and evolution, helping to elucidate how the polygynous form came to predominate. The USDA, too, brought the insect into the laboratory.

In the 1990s, this research seemed to reach fruition when entomologists pioneered a new way to control the ant—not eradicating, but rather taming it. About two decades earlier, USDA entomologists working in Brazil found several species of phorid flies—gnat-sized insects— that laid their eggs on *Solenopsis invicta*. The maggots developed, sucking vital fluids from their ant hosts, eventually decapitating them. The phorids killed less than 5 percent of the ants, however, and so seemed an unlikely weapon against the abundant pests. A few years later, researchers at the University of Texas reignited interest in the parasites when they found that the imported fire ant cowered in the presence of the flies, reducing the scope of its foraging and showing a surprising vulnerability to competition from other ants. Texas entomologist Lawrence Gilbert pressed for more research; a quail hunter donated $10,000; and Sanford Porter, a post-doctoral researcher working with Gilbert, joined the USDA, carrying with him an interest in the flies.

Entomologists viewed the possibility of using the flies to control the ant with cautious optimism. Although biological control is often a more elegant solution to insect problems than insecticides, it is not without its dangers: the effect of releasing an alien plant or animal into the environment ripples outward, and consequences are not always predictable.[93] In laboratory tests, phorid flies preferred to attack *Solenopsis invicta*, but they also attacked native ants.[94] There was a small but not insignificant possibility that the flies might further deci-

92. On Buren's job and its goals, see W. Whitcomb to D. Pimentel, 1 April 1969, W. H. Whitcomb Tall Timbers Conference file, box 5, David Pimentel papers, 21/23/1453, Cornell University Archives; W. F. Buren to W. S. Creighton, 2 July 1971 and 2 February 1973, Creighton papers.

93. On the risks of environmental control, see F. G. Howarth, "Environmental Impacts of Classical Biological Control," *Annual Review of Entomology* 36 (1991): 485–509; D. J. Greathead, "Benefits and Risks of Classical Biological Control," *Plant and Microbial Biotechnology Research Series* 4 (1995): 53–63; J. A. Lockwood, "The Ethics of Biological Control: Understanding the Moral Implications of Our Most Powerful Ecological Technology," *Agriculture and Human Values* 13 (1996): 2–19; J. A. Lockwood, "Competing Values and Moral Imperatives: An Overview of Ethical Issues in Biological Control," *Agriculture and Human Values* 14 (1997): 205–10.

94. S. Porter, "Host Specificity and Risk Assessment of Releasing the Decapitating Fly *Pseudacteon curvatus* as a Classical Biological Control Agent for Imported Fire Ants," *Biological Conservation* 106 (2000): 35–47.

mate populations of *Solenopsis geminata* and *Solenopsis xyloni*.[95] Porter, though, argued that the flies attacked the native ants infrequently, the ants were populous enough to survive some parasitism, and the need to control the imported fire ant was so great that the possible reward of releasing the flies was worth the potential risks.[96]

In 1995, scientists released phorids into the Texas environment.[97] The first test results were a failure, as harsh weather wiped out the parasite population before the fly could inflict much damage. Fire ant researchers then realized that they needed many generations of the fly in the wild at the same time: adults flitting from place to place as well as maggots squirming in the fire ants' nests. Weather might destroy a cohort of adults, but the young would be protected from the environment. "While Brazil also may have episodes of harsh weather which might kill adult populations, any losses will soon be replaced by adults eclosing from pupae," Gilbert explained. "The presence of eggs, larvae and pupae in the natural population acts as a buffer against short-term disaster."[98] The scientists built tropical greenhouses to raise the fly, creating a steady, diverse, and stable population.[99] Only because they remade parts of South America in North America and controlled the environment could they hope for success. The control of nature may have characterized the Neanderthal age of biology, but it was also needed in the space age of biology.

"As far as I know," Gilbert said in 1996, "there has never been a successful biocontrol of a pest ant. My colleagues and I would like to change that dismal fact."[100] "Control" had become the term of art, not

95. On this point, see J. Van Driesche and R. Van Driesche, "Weighing Urgency against Uncertainty: The Conundrum of Biocontrol," *Conservation in Practice* 4 (2003): 24–25.

96. S. Porter, "Host Specificity and Risk Assessment of Releasing the Decapitating Fly *Pseudacteon curvatus* as a Classical Biological Control Agent for Imported Fire Ants," *Biological Conservation* 106 (2000): 35–47.

97. On this history, see L. E. Gilbert, "Prospects of Controlling Fire Ants with Parasitoid Flies: The Perspective from Research Based at Brackenridge Field Laboratory," 23 March 1996, available at www.utexas.edu/research/bfl/research/gilbert/prospects.html (accessed 28 August 2002).

98. Ibid.

99. For overviews of the work, see S. D. Porter, D. F. Williams, and R. S. Patterson, "Rearing the Decapitating fly *Pseudacteon tricuspis* (Diptera: Phoridae) in Imported Fire Ants (Hymenoptera: Formicidae) from the United States," *Journal of Economic Entomology* 90 (1997): 135–38; M. R. Orr, S. H. Seike, W. W. Benson, and L. E. Gilbert, "Flies Suppress Fire Ants," *Nature* 373 (1995): 292–93; and S. D. Porter, M. A. Pesquero, S. Campiolo, and H. G. Fowler, "Growth and Development of *Pseudacteon* Phorid Fly Maggots (Diptera: Phoridae) in the Heads of *Solenopsis* Fire Ant Workers (Hymenoptera: Formicidae)," *Environmental Entomology* 24 (1995): 475–79.

100. L. E. Gilbert, "Prospects of Controlling Fire Ants with Parasitoid Flies: The Perspective from Research Based at Brackenridge Field Laboratory," 23 March 1996, available at www.utexas.edu/research/bfl/research/gilbert/prospects.html (accessed 28 August 2002).

eradication. Pesticides traditionally used in residential areas to control mosquitoes and other pests also killed the flies, and so the parasites could not become established in urban and suburban settings; for this reason, the phorids would be unable to exterminate the ant.[101] In rural areas, however, the ant's polygyny made them an easy target. As the number of ants increased, *Solenopsis invicta* was easier to find, and the number of parasites increased, too. Research entomologists imagined islands of native ants, fortified by the presence of the parasites, emerging out of the sea of *Solenopsis invicta*, beachheads in a battle that pitted insect against insect.[102] Eventually, they hoped, the natives would regain their dominance and the imported fire ant would slip into the background, another member of a diverse ant community. As the twentieth century concluded, results of the experiments remained inconclusive, but the USDA exuded confidence. "We're excited as heck," said one entomologist.[103] Sanford Porter imagined that releasing even more predators, parasites, and diseases would further erode the imported fire ant's competitive advantage. "We hope they will be a three- or four-edged sword," he said in 2002.[104]

Lessons

By the end of the century, and despite the backlash of the 1980s, environmentalism had a strong grip on America. A Gallup poll of 1,004 adults in April 2000 found that 66 percent of Americans supported environmental protection. (Respondents ranked education, health care, crime, social security, family values, the economy, and gun control more pressing issues for the president; tax cuts, foreign affairs, abortion, and campaign finance reform elicited less interest.)[105] Younger people were less likely than older to identify themselves as environmentalists, suggesting that the cultural upheaval of the 1960s may have had a greater impact on the development of environmentalism than historian Samuel Hays would allow (see chapter 3). But still, about half of Americans age thirty or older called themselves environmentalists, evidence that environmentalism was not just an outgrowth

101. C. McIntosh, "Fly vs. Ant: Auburn Scientists Use Phorid Fly to Battle Fire Ants," *Birmingham Post-Herald*, 18 May 2002, C1.

102. L. E. Gilbert, "Prospects of Controlling Fire Ants with Parasitoid Flies: The Perspective from Research Based at Brackenridge Field Laboratory," 23 March 1996, available at www.utexas.edu/research/bfl/research/gilbert/prospects.html (accessed 28 August 2002).

103. B. Reid, "Fire Ants: Creatures Biggest Pest in South," *Clarion Ledger*, 9 June 1996, 1A.

104. C. McIntosh, "Fly vs. Ant: Auburn Scientists Use Phorid Fly to Battle Fire Ants," *Birmingham Post-Herald*, 18 May 2002, C1.

105. G. Martin, "Earth Day Report Card—We Still Care, Sort of," *San Francisco Chronicle*, 22 April 2000, A1.

of the student movement and counterculture revolution of the 1960s, but reflected fundamental changes in American society and economy.[106] Increasingly, nature was not seen as a set of resources to be exploited, but as a complexly integrated nexus of relations on which human life depended. The key to the good life was protecting its beauty, preserving its health, and maintaining its permanence.

Disputes about how to best make a livable world persisted, however. Some continued to confuse environmentalism with conservation; some continued to focus on the punitive measures pioneered by the EPA in the 1970s; some pushed to connect environmentalism to civil rights; some looked askance at government regulation. Others experimented with what University of Wisconsin professor of public and environmental affairs Michael E. Kraft identified as the "third wave" of environmentalism, sustainable development.[107] Following the enactment of environmental laws in the 1960s and 1970s and retrenchment in the 1980s, sustainable development advocates tried to broker the differences between the two eras, seeking ways to protect the environment while simultaneously encouraging economic development. In 1992, E. O. Wilson, having become deeply involved with the environmental movement during the previous decade, described the essence of what he called the "new environmentalism": "Except in pockets of ignorance and malice, there is no longer an ideological war between conservationists and developers. Both share the perception that health and prosperity decline in a deteriorating environment. . . . The race is on to develop methods, to draw more income from the wildlands without killing them, and so to give the invisible hand of free-market economics a green thumb."[108]

The sustainable development movement, like integrated pest management, to which it was sometimes linked, had deep roots. Clarence Cottam and Robert Rudd, for example, had argued early on for the need to connect environmental protection and economic production. In the 1940s, Cottam helped to design a wildlife refuge for the Tennessee Valley Authority. He argued that farmers should not pay the government to work the land, but should instead leave a percentage

106. Ibid. The generation gap broke down as follows: 37 percent of people age eighteen to twenty-nine rated themselves environmentalists, while 47 percent of those age thirty to forty-nine, 49 percent of those fifty to sixty, and 57 percent of those age sixty-five and older did.

107. M. E. Kraft, "U.S. Environmental Policy and Politics: From the 1960s to the 1990s," *Journal of Policy History* 12 (2000): 33–36.

108. E. O. Wilson, *The Diversity of Life* (Cambridge: Harvard University Press, 1992), 282–83. The ellipse collapses a longer discussion, but the gist of it is maintained. For Wilson's developing interest in environmental problems, see E. O. Wilson, *Naturalist* (Washington, D.C.: Island Press, 1994), 354–55.

of their crops in the field, unharvested, as a way to attract birds.[109] Production and protection could go hand-in-hand. Rudd offered a similar proposal for solving the pesticide problem. Simply stopping the use of insecticides, he said, would not help, since the landscape favored the continued spread of insects and made the chemicals necessary. The way forward was to change farming practices, thus minimizing the effect of insect outbreaks while maintaining production at a high enough level to provide cheap, healthy food for Americans.[110] "The fields, forests, ranges, and waters are living systems capable of change," he wrote. "With wise and cautious manipulation these producing environments will continue to supply our needs and pleasures."[111] And Aldo Leopold, although he became skeptical of the conservation ideals he had once held, never completely broke free of them; instead, he struggled toward a reconciliation of conservation and environmentalism. We are tinkerers, he said, working on the complex machinery of nature and even as we endeavor to protect nature, we change it, to make a world in which we can thrive.[112]

It is no coincidence that all three of these men were wildlife biologists; wildlife biology, as historians Samuel Hays and Thomas Dunlap note, provided a bridge between the gospel of efficiency and the environmental movement, and Cottam, Rudd, and Leopold were active during the time that bridge was under construction, drawing inspiration from the conservation ideal of managing the land but also acknowledging that the land had an integrity of its own that could not be altered without consequence, both good and ill.[113] The historian Donald Worster has criticized Leopold's mixed message, complaining that he continued to see nature too much as a machine that could be manipulated and did not sufficiently admit that humans needed to change their behavior to fit into nature's economy.[114] But, I would

109. C. Cottam, "The Role of Impoundments in Post-War Planning for Waterfowl," *Transactions of the North American Wildlife Conference* 9 (1944): 288–95; H. C. Throup, "Clarence Cottam: Conservationist: The Welder Years" (Ph.D. diss., Brigham Young University, 1983), 107–9.

110. R. Rudd, *Pesticides and the Living Landscape* (Madison: University of Wisconsin Press, 1964), 288.

111. Ibid., 6.

112. A. Leopold, *A Sand County Almanac* (New York: Ballantine Books, [1949] 1970), 190.

113. S. P. Hays, *Beauty, Health, and Permanence: Environmental Politics in the United States, 1955–1985* (New York: Cambridge University Press, 1987), 21; T. R. Dunlap, *Saving America's Wildlife: Ecology and the American Mind, 1850–1990* (Princeton: Princeton University Press, 1988), 98–105.

114. D. Worster, *Nature's Economy: A History of Ecological Ideas* (New York: Cambridge University Press [1977] 1996), 289–90.

argue, it is exactly this mixture that makes Leopold's ideas relevant today.

As best I can determine, though, neither Leopold, Rudd, nor Cottam influenced the new environmentalism of the 1990s. It developed according to its own logic. The goal was to initiate a new industrial revolution using new technologies and new practices that repaid entrepreneurs without destroying the natural system on which our lives depend. This was the era of the eco-economy, of natural capitalism, of the new economy of nature.[115] Daniel Janzen, for example, son of the U.S. Fish and Wildlife Service (USFWS) official who served in the fire ant wars, linked the protection of forests in Costa Rica to the economic development of the area. He hired locals to suppress fires, their paychecks proof of the forest's worth. He invited tourists, and charged the fifty thousand annual visitors six dollars at the gate. Janzen also inked a deal with pharmaceutical giant Merck, giving the corporation two years of exclusive rights to chemical extracts from park organisms in return for $1 million and royalties if Merck successfully developed drugs from the chemicals. In 1998, he forged a partnership with a local juice plant. The company produced three hundred pounds of pulp each day and needed somewhere to dispose of it. Janzen agreed to take the pulp, then dumped it on ground infested with imported grasses that he wanted to turn into forest. The pulp smothered the unwanted plants and fertilized the natives: out of the precipitate of corporate capitalism arose a new forest. Humans created nature, and profited from it.[116] "You invest in nature," Janzen said. "Use it, think of it as having an owner and a future, as producing goods and services for the country, rather than as a frontier to exploit or destroy."[117]

Even the Environmental Defense Fund altered its "sue the bastards" style and sought ways to combine environmentalism and growth. In the mid-1990s, the EDF sold the USFWS on its "safe harbors" program, one of the few environmental innovations to receive accolades from

115. These figure as the titles for three books: L. R. Brown, *Eco-economy: Building an Economy for the Earth* (New York: Norton, 2001); P. Hawken, A. Lovins, and L. H. Lovins, *Natural Capitalism: Creating the Next Industrial Revolution* (Boston: Little, Brown, 1999); and G. C. Daily, and K. Ellison, *The New Economy of Nature: The Quest to Make Conservation Profitable* (Washington, D.C.: Island Press, 2002). See also R. Costanza and R. V. O'Neill, eds., "Ecological Economics," *Ecological Applications* 6 (1996): 975–1034.

116. On Janzen's work, see W. Allen, *Green Phoenix: Restoring the Tropical Forests of Guanacaste, Costa Rica* (New York: Oxford University Press, 2001); G. C. Daily and K. Ellison, *The New Economy of Nature: The Quest to Make Conservation Profitable* (Washington, D.C.: Island Press, 2002), 165–88.

117. G. C. Daily and K. Ellison, *The New Economy of Nature: The Quest to Make Conservation Profitable* (Washington, D.C.: Island Press, 2002), 171.

both the National Audubon Society and the American Farm Bureau.[118] In this program, the fish and wildlife service encouraged landowners to make their property attractive to endangered species by promising that if endangered species did take up residence, the government would not burden the landowners with costly new regulations. "There are a lot of landowners . . . who are willing to lay out the welcome mat for endangered species on their land if the reward for doing so is no additional restrictions," said EDF lawyer Michael Bean, who helped to created the program.[119] In years past, many property owners had worried that if an endangered species found its way onto their land, they would be prevented from developing their land as they saw fit, and so they made their property unattractive to the organisms. The program allowed the landowner to continue to use his or her property while also working to protect the environment.[120] Promoting a similar idea to Congress in 2001, California rancher Ralph Grossi recommended, "The next farm bill should be designed to help farmers provide environmental quality and reward them when they do so."[121]

Some, of course, opposed this trend, eulogizing the lost activism for a romanticized Nature of earlier years.[122] In his widely read 1996 book *The End of Nature*, journalist Bill McKibben decried the new environmentalism as surrender. "This intended rallying cry," he wrote, "depresses me more deeply than I can say. That is our destiny? To be 'caretakers' of a managed world, 'custodians' of all life? For that job security we will trade the mystery of the natural world, the pungent mystery of our own lives and of a world bursting with exuberant creation?"[123] McKibben, though, misrepresents the trade. Seeing the human hand in nature does not mean that nature is gone. Even the most domesticated nature can be elusive, as a gardener of any humble, worked-over patch of land can attest.[124] The imported fire ant, transported and sustained by human practices,

118. On the program, its development, its supporters, and its limitations, see D. S. Wilcove, M. J. Bean, R. Bonnie, and M. McMillan, "Rebuilding the Ark: Toward a More Effective Endangered Species Act for Private Land," 5 December 1996, available at www.environmentaldefense. org/documents/483_Rebuilding%20the%20Ark%2Ehtm (accessed 17 October 2002).

119. M. Lee, "Program Unites Farmers, Endangered Species," *Sacramento Bee*, 13 May 2003, D1.

120. See also D. L. Jackson and L. L. Jackson, *The Farm as Natural Habitat: Reconnecting Food Systems with Ecosystems* (Washington, D.C.: Island Press, 2002).

121. T. Holt, "Future Farmers of Ecology," *Sacramento Bee*, 25 May 2003, E3.

122. For example, M. Dowie, *Losing Ground: American Environmentalism at the Close of the Twentieth Century* (Cambridge: MIT Press, 1995); J. Terborgh, *Requiem for Nature* (Washington, D.C.: Island Press, 1999).

123. Bill McKibben, *The End of Nature* (New York: Random House, 1989), 214.

124. Compare D. G. Schueler, *A Handmade Wilderness* (Boston: Houghton Mifflin, 1996); D. Ackerman, *Cultivating Delight: A Natural History of My Garden* (New York: Harper Perennial, 2002).

contained by cultural categories, has continued to surprise, revealing characteristics that no one expected. Nature remains mysterious. And managing the world is not a capitulation to the forces of evil, but a responsibility. Humans have shaped the world, even those parts that seem natural. The South, the ant's defenders thought, operated according to natural rules that were independent from human influence. But Americans had built the landscape, and leaving the ant alone allowed it to wreak havoc. Failing to manage our creations is an abdication of responsibility.

I admit to being drawn to this view of the relationship between nature and human life. It is not the one I held when I began this study of the fire ant wars. Then, I was closer to McKibben than to Janzen. When I took Janzen's class on humans and the environment at the University of Pennsylvania—in which I wrote an essay on the fire ant that would ultimately become this book—I was uncomfortable with his views. Over time, however, as I further researched the story of the imported fire ant, I came to see the wisdom of his views, and I think that they embody some of the chief lessons to be learned from tracing this part of nature. Nature existed long before humans, and will continue to exist long after us. Nature is everything, everywhere, at all times, and even the nothingness that follows the collapse of the universe or the dead cold left in the wake of its infinite expansion would still be natural. But, today, nature is not just something beyond us. It is something that we create and recreate through our practices. Biological control, cultural control, chemical control—they all require altering nature, shaping and changing it, as the greenhouses built for the phorid flies proved. "The fundamental question," writes philosopher Steven Vogel, "is not 'how ought we to interact with nature?' as though [nature] were something independent of us and given from all eternity. . . . Rather the question to be asked is more correctly, 'What ought the environment to be?'"[125] Do we want an agricultural landscape that requires the use of insecticides for its maintenance? Do we want an alien insect that disrupts the southern ecosystem, destroys crops, and harasses people? And, if not, what are we willing to do to change the situation, keeping in mind that we created the situation in the first place to address some other set of problems and that the changes might result in unknown consequences. How can we produce the just, beautiful, and strong world that exists now only in our minds?

My enthusiasm for this view, though, is tempered by two other lessons gleaned from the fire ant wars. First, they teach that ideas about

125. S. Vogel, *Against Nature: The Concept of Nature in Critical Theory* (Albany: State University of New York Press, 1996), 168.

nature are always bound to particular moments in history, dependent on the structure of nature and the organization of society at a certain time.[126] Entomologists with the PPC, inheritors of the gospel of efficiency, saw nature as something that could be managed completely. Environmentalists, riding America's post–World War II changes in demography and economy, saw nature as a fragile web of relationships. Fifteen hundred years ago, neither of these views would have made any sense; and fifteen hundred—or one hundred, or fifteen—years in the future, this third wave of environmentalism may seem outmoded. Things change. There is no path back to the Garden of Eden.[127]

Instead, we must make our way in the world as best we can; but what we imagine, and what we hope to accomplish, will rarely match reality. This, I think, is the second cautionary lesson of the fire ant wars. As humans have grown increasingly powerful, our actions have become inextricably bound with nature, but we do not yet know nature. The high rate of species extinction proves that we do not even know what Aldo Leopold called "the first principle": keep all the parts.[128] And after we learn that lesson there are many more. Still, despite this ignorance, we must continue to work with nature, and in the process recreate it. We can imagine the consequences of some of our actions; we have not the slightest clue about others. In many cases we do not even know what we do not know. We shape nature, but do not always control what we make: replacing insecticides, for example, with biological control agents solves some problems, but cause others, as biological control, too, can go awry. So, even as we might be optimistic about the future and imagine new ways of relating to the world, the fire ant wars teaches us to be humble and vigilant, to expect the unexpected, and to know that tragedy and triumph usually come together.[129] The fire ant wars do not tutor despair, but an acknowledgment of our limits as humans, even as we strive to overcome them. "We shall never achieve harmony with the land, any more than we shall achieve absolute justice or liberty for people," Leopold wrote. "In these higher aspirations the important thing is not to achieve, but to strive."[130]

126. W. Cronon, "The Trouble with Wilderness; or, Getting Back to the Wrong Nature," in *Uncommon Ground: Rethinking the Human Place in Nature*, ed. W. Cronon (New York: Norton, 1996), 69–90.

127. Compare E. Eisenberg, *The Ecology of Eden: An Inquiry into the Dream of Paradise and a New Vision of our Role in Nature* (New York: Vintage Books, 1998).

128. A. Leopold, *A Sand County Almanac* (New York: Ballantine Books, [1949] 1970), 190.

129. On this point, see W. Cronon, "The Uses of Environmental History," *Environmental History Review* 18 (1994): 15.

130. A. Leopold, *A Sand County Almanac* (New York: Ballantine Books, [1949] 1970), 210.

References

Archival Collections

Agricultural Research Service papers, RG 310, National Archives and Records Administration, College Park, Md.

Alabama Department of Agriculture and Industry papers, Alabama Department of Archives and History, Montgomery, Ala.

Alabama Department of Conservation, director's administrative papers, Alabama Department of Archives and History, Montgomery, Ala.

Frank Selma Arant papers, RG 789, Auburn University.

Bird and Mammal Laboratories, U.S. Fish and Wildlife Service, *ca.* 1885–1971, Records, Record Unit 7171, Smithsonian Institution Archives, Washington, D.C.

Murray S. Blum papers, in author's possession.

Murray S. Blum–William Steel Creighton correspondence, City College of New York Archives.

James C. Bradley papers, 21/23/1717, Division of Rare and Manuscript Collections, Cornell University Library.

Frank Morton Carpenter papers, HUG 4264.5, Harvard University Archives.

Rachel Carson papers, Yale Collection of American Literature, 46, Beinecke Rare Book and Manuscript Library, Yale University.

Arthur Charles Cole papers, an uncatalogued part of the E. O. Wilson papers, Library of Congress, Washington, D.C.

William Steel Creighton papers, an uncatalogued part of the E. O. Wilson papers, Library of Congress, Washington, D.C.

Philip Jackson Darlington papers, HUG (FP) 75.10, Harvard University Archives.

Ralph B. Draughon papers, RG 107, Auburn University.

Entomological Society of America papers, MS-488, Iowa State University Archives, Ames, Iowa.

Environmental Defense Fund papers, MS 232, State University of New York, Stony Brook.

Extension Entomology papers, RG 842, Auburn University.

Fire Ant Correspondence, series 2012, Mississippi Department of Agriculture and Commerce papers, Mississippi Department of Archives and History, Jackson, Miss.

Warren E. Hinds collection, RG 615, Auburn University.

National Academy of Sciences (Biology and Agriculture) papers, National Academy of Sciences Archives, Washington, D.C.

National Audubon Society Records, Manuscripts and Archives Division, Astor, Lenox and Tilden Foundations, New York Public Library, New York, N.Y.

William Nichols papers, RG 194 (84-1), Auburn University.

David Pimentel papers, 21/23/1453, Division of Rare and Manuscript Collections, Cornell University Library.

Plant Pest Control Division papers, RG 463, National Archives and Records Administration, College Park, Md. [Some citations only refer to the year. The relevant boxes can be ordered at the National Archives using the Record Group, subject, and year.]

Walter Rosene papers, unprocessed, Auburn University.

Paul B. Sears papers, 663, Sterling Library, Yale University Archives.

Secretary of Agriculture papers, RG 16, National Archives and Records Administration, College Park, Md.

Secretary of the Interior papers, RG 48, National Archives and Records Administration, College Park, Md.

Systematic Entomology Laboratory, U.S. Department of Agriculture, 1797–1988 and undated. Photographs and Biographical Information, Record Unit 7323, Smithsonian Institution Archives, Washington, D.C.

U.S. Fish and Wildlife papers, RG 22, National Archives and Records Administration, College Park, Md.

John Whitaker Subject Files, White House Central Files, Richard M. Nixon papers, National Archives and Records Administration, College Park, Md.

William Morton Wheeler papers, Ernst Mayr Library, Museum of Comparative Zoology Archives, Harvard University.

William Morton Wheeler papers, HUG (FP) 87.10, Harvard University Archives.

Edward O. Wilson papers, uncatalogued, Library of Congress, Washington, D.C.

Books and Articles

Ackerman, D. *Cultivating Delight: A Natural History of My Garden.* New York: Harper Perennial, 2002.

Adkins, H. G. "The Imported Fire Ant in the Southern United States." *Annals of the Association of American Geographers* 60 (1970): 578–92.

Adler, L. K., and T. G. Paterson. "Red Fascism: The Merger of Nazi Germany and Soviet Russia in the American Image of Totalitarianism, 1930's to 1950's." *American Historical Review* 75 (1970): 1046–64.

Alabama: A Guide to the Deep South. New York: R. R. Smith, 1941.

Allen, W. *Green Phoenix: Restoring the Tropical Forests of Guanacaste, Costa Rica.* New York: Oxford University Press, 2001.

Allred, R. "Catharsis, Revision, and Re-Enactment: Negotiating the Meaning of the American Civil War." *Journal of American Culture* 19 (1996): 1–13.

Andrews, R. N. L. *Managing the Environment, Managing Ourselves: A History of American Environmental Policy.* New Haven: Yale University Press, 1999.

Arant, F. S. *The Status of Game Birds and Mammals in Alabama.* Wetumpka, Ala: Wetumpka Printing, 1939.

Babcock, H. *I Don't Want to Shoot an Elephant.* New York: Holt, Rinehart and Winston, 1958.

———. *My Health Is Better in November.* Columbia: University of South Carolina Press, 1947.

Baker, G. L. *Century of Service: The First 100 Years of the United States Department of Agriculture.* Washington, D.C.: USDA, 1963.

Balogh, B. *Chain Reaction: Expert Debate and Public Participation in American Commercial Nuclear Power, 1945–1975.* Cambridge: Cambridge University Press, 1991.

Banks, W. A., B. M. Glancey, C. E. Stringer, D. P. Jouvenaz, C. S. Lofgren, and D. E. Weidhaas. "Imported Fire Ants: Eradication Trials with Mirex Bait." *Journal of Economic Entomology* 66 (1973): 785–89.

Bates, H. W. *The Naturalist on the River Amazons.* London: John Murray, [1863] 1892.

Battenfield, S. L., ed. *Proceedings of the Symposium on the Imported Fire Ant.* Washington, D.C.: Government Printing Office, 1982.

Benson, E. T. *Cross Fire: The Eight Years with Eisenhower.* Garden City, N.Y.: Doubleday, 1962.

———. *Freedom to Farm.* Garden City, N.Y.: Doubleday, 1960.

Berger, S. R. *Dollar Harvest: The Story of the Farm Bureau.* Lexington, Mass.: Heath Lexington Books, 1971.

Bloor, D. *Knowledge and Social Imagery.* Boston: Routledge and Keegan Paul, 1976.

Boffey, P. M. *The Brain Bank of America: An Inquiry into the Politics of Science.* New York: McGraw-Hill, 1975.

Bonettos, A. A., and I. R. Wais. "Southern South American Streams and Rivers." In *River and Stream Ecosystems*, ed. C. E. Cushing, K. W. Cummins, and G. W. Minshall, 257–93. Vol. 22 of *Ecosystems of the World*, ed. D. W. Goodall. Amsterdam: Elsevier, 1995.

Bosso, C. J. *Pesticides and Politics: The Life Cycle of a Public Issue*. Pittsburgh: University of Pittsburgh Press, 1987.

Bowler, P. J. *Life's Splendid Drama: Evolutionary Biology and the Reconstruction of Life's Ancestry 1860–1940*. Chicago: University of Chicago Press, 1996.

Brown, L. R. *Eco-economy: Building an Economy for the Earth*. New York: Norton, 2001.

Buhs, J. B. "Building on Bedrock: William Steel Creighton and the Reformation of Ant Systematics." *Journal of the History of Biology* 33 (2000): 55–64.

———. "Dead Cows on a Georgia Field: Mapping the Cultural Landscape of the Post–World War II American Pesticide Controversies." *Environmental History* 7 (2002): 99–121.

———. "The Fire Ant Wars: Nature and Science in the Pesticide Controversies of the Late Twentieth Century." *Isis* 93 (2002): 377–400.

Buren, W. F. "Revisionary Studies on the Taxonomy of the Imported Fire Ants." *Journal of the Georgia Entomological Society* 7 (1972): 1–26.

Buren, W. F., G. E. Allen, W. H. Whitcomb, F. E. Lennartz, and R. N. Williams. "Zoogeography of the Imported Fire Ants." *Journal of the New York Entomological Society* 82 (1974): 113–24.

Carson, R. *Silent Spring*. Boston: Houghton Mifflin, [1962] 1994.

Carter, D. T. *George Wallace, Richard Nixon, and the Transformation of American Politics*. Waco, Tex: Markham Press Fund, 1992.

Clark, J. F. M. " 'The Ants Were Duly Visited': Making Sense of John Lubbock, Scientific Naturalism, and the Senses of Social Insects." *British Journal for the History of Science* 30 (1997): 151–76.

Clarke, J. N., and D. C. McCool. *Staking Out the Terrain: Power and Performance among Natural Resource Agencies*. Albany: State University of New York Press, 1996.

Committee on the Future Role of Pesticides in U.S. Agriculture. *The Future Role of Pesticides in US Agriculture*. Washington, D.C., National Academy Press, 2000.

Cowan, F. *Curious Facts in the History of Insects*. Philadelphia: Lippincott, 1865.

Cowdrey, A. E. *This Land, This South: An Environmental History*. Lexington: University Press of Kentucky, 1996.

Creighton, W. S. "The New World Species of the Genus *Solenopsis* (Hymenoptera: Formicidae)." *Proceedings of the American Academy of Arts and Sciences* 66 (1930): 39–151.

Cronon, W. "Modes of Prophecy and Production: Placing Nature in History." *Journal of American History* 76 (1990): 1122–31.

———. "A Place for Stories: Nature, History, and Narrative." *Journal of American History* 78 (1992): 1347–76.

———. "The Uses of Environmental History." *Environmental History Review* 17 (1993): 1–22

Cronon, W., ed. *Uncommon Ground: Rethinking the Human Place in Nature*. New York: Norton, 1996.

Dahlstein, D. L., and R. Garcia, eds. *Eradication of Exotic Pests*. New Haven: Yale University Press, 1989.

Daily, G. C., and K. Ellison. *The New Economy of Nature: The Quest to Make Conservation Profitable*. Washington, D.C.: Island Press, 2002.

Daniel, P. *Breaking the Land: The Transformation of Cotton, Tobacco, and Rice Cultures since 1880*. Urbana: University of Illinois Press, 1985.

———. *Lost Revolutions: The South in the 1950s*. Chapel Hill: University of North Carolina Press, 2000.

———. "A Rogue Bureaucracy: The USDA Fire Ant Campaigns of the Late 1950s." *Agricultural History* 64 (1990): 99–114.

———. *Standing at the Crossroads: Southern Life in the Twentieth Century*. New York: Hill and Wang, 1986.

Devine, R. S. *Alien Invasion: America's Battle with Non-Native Animals and Plants*. Washington, D.C.: National Geographic Society, 1998.

Dos Passos, J. *State of the Nation*. Boston: Houghton Mifflin, 1944.

Douglas, G., R. H. Jones, and P. Henegar, eds. *The History of the Southern Nurserymen's Association, Inc., 1899–1974*. Nashville, Tenn: Southern Nurserymen's Association, 1974.

Dowie, M. *Losing Ground: American Environmentalism at the Close of the Twentieth Century*. Cambridge: MIT Press, 1995.

Dunlap, T. R. *DDT: Scientists, Citizens, and Public Policy*. Princeton: Princeton University Press, 1981.

———. *Saving America's Wildlife: Ecology and the American Mind, 1850–1990*. Princeton: Princeton University Press, 1988.

Dupree, A. H. *Science in the Federal Government: A History of Policies and Activities*. Baltimore: Johns Hopkins University Press, [1957] 1986.

Egoroff, P. P. *Argentina's Agricultural Exports during World War II*. Palo Alto: Food Research Institute, Stanford University, 1945.

Eisenberg, E. *The Ecology of Eden: An Inquiry into the Dream of Paradise and a New Vision of Our Role in Nature*. New York: Vintage Books, 1998.

Elton, C. S. *The Ecology of Invasions by Animals and Plants*. London: Chapman and Hall, [1958], 1977.

Evans, E. P. *The Criminal Prosecution and Capital Punishment of Animals*. London: Faber and Faber, [1906] 1978.

Evans, M. A., and H. E. Evans. *William Morton Wheeler, Biologist*. Cambridge: Harvard University Press, 1970.

Fite, G. C. *Cotton Fields No More: Southern Agriculture, 1865–1980*. Lexington: University Press of Kentucky, 1984.

Fitzgerald, D. *Every Farm a Factory: The Industrial Ideal in American Agriculture*. New Haven: Yale University Press, 2003.

Flader, S. *Thinking Like a Mountain: Aldo Leopold and the Evolution of an Ecological Attitude toward Deer, Wolves, and Forests*. Columbia: University of Missouri Press, 1974.

Flippen, J. B. *Nixon and the Environment*. Albuquerque: University of New Mexico Press, 2000.

———. "Pests, Pollution, and Politics: The Nixon Administration's Pesticide Policy." *Agricultural History* 71 (1997): 442–56.

Fortune, I. "The Biology and Control of *Solenopsis saevissima* variety *richteri* Forel." Master's thesis, Mississippi State College, 1948.

Freeman, M., ed. *Always, Rachel: The Letters of Rachel Carson and Dorothy Freeman, 1952–1964*. Boston: Beacon Press, 1995.

Gantham, D. W. *The South in Modern America: A Region at Odds*. New York: Harper Collins Publishers, 1994.

George, J. L. *The Pesticide Problem*. New York: Conservation Foundation, 1957.

———. *The Program to Eradicate the Imported Fire Ant*. New York: Conservation Foundation, 1958.

Gilbert, L. E. "Prospects of Controlling Fire Ants with Parasitoid Flies: The Perspective from Research Based at Brackenridge Field Laboratory." 23 March 1996, available at www.utexas.edu/research/bfl/research/gilbert/prospects.html, accessed 28 August 2002.

Golinski, J. *Making Natural Knowledge: Constructivism and the History of Science.* New York: Cambridge University Press, 1998.

Gottlieb, R. *Forcing the Spring: The Transformation of the Environmental Movement.* Washington, D.C.: Island Press, 1993.

Graham, F. *The Audubon Ark.* New York: Knopf, 1990.

———. *Since Silent Spring.* Boston: Houghton Mifflin, 1970.

Hamilton, D. E. "Building the Associative State: The Department of Agriculture and American State-Building." *Agricultural History* 64 (1990): 207–18.

Haskins, C. P. *Of Ants and Men.* New York: Prentice-Hall, 1939.

———. *Of Societies and Men.* New York: Viking, 1951.

Hawken, P., A. Lovins, and L. H. Lovins. *Natural Capitalism: Creating the Next Industrial Revolution.* Boston: Little, Brown, 1999.

Hayes, W. J. "Pesticides in Relation to Public Health." *Annual Review of Entomology* 5 (1960): 379–404.

Hays, S. P. *Beauty, Health, and Permanence: Environmental Politics in the United States, 1955–1985.* New York: Cambridge University Press, 1987.

———. *Conservation and the Gospel of Efficiency: The Progressive Conservation Movement, 1890–1920.* Cambridge: Harvard University Press, 1959.

Hays, S. "The Food Habits of the Imported Fire Ant, *Solenopsis saevissima richteri* Forel, and Poison Baits for Its Control." Master's thesis, Alabama Polytechnic Institute, 1958.

Helms, J. D. "Just Lookin' for a Home: The Cotton Boll Weevil and the South." Ph.D. diss., University of Florida, 1977.

Hogan, M. *A Cross of Iron: Harry S. Truman and the Origins of the National Security State, 1945–1954.* New York: Cambridge University Press, 1998.

Hölldobler, B., and E. O. Wilson. *Journey to the Ants: A Story of Scientific Exploration.* Cambridge: Harvard University Press, 1994.

Hooks, G. "From an Autonomous to a Captured State Agency: The Decline of the New Deal in Agriculture." *American Sociological Review* 55 (1990): 29–43.

Howarth, F. G. "Environmental Impacts of Classical Biological Control." *Annual Review of Entomology* 36 (1991): 485–509.

Hoyt, E. *The Earth Dwellers: Adventures in the Land of Ants.* New York: Touchstone Books, 1996.

Huxley, J. *Ants.* New York: Jonathan Cape and Harrison Smith, 1930.

Jackson D. L., and L. L. Jackson. *The Farm as Natural Habitat: Reconnecting Food Systems with Ecosystems.* Washington, D.C.: Island Press, 2002.

Johnson, A. S. "Antagonistic Relationships between Ants and Wildlife." Master's thesis, Auburn University, 1962.

Kaiser, K. L. E. "The Rise and Fall of Mirex." *Environmental Science and Technology* 12 (1978): 520–28.

Kirby, J. T. *Rural Worlds Lost: The American South 1920–1960.* Baton Rouge: Louisiana State University Press, 1987.

Kogan, M. "Integrated Pest Management: Historical Perspectives and Contemporary Developments." *Annual Review of Entomology* 43 (1998): 243–70.

Kohler, R. E. *Lords of the Fly: Drosophila Genetics and the Experimental Life.* Chicago: University of Chicago Press, 1994.

Komarek, E. V. "Comments on the 'Fire Ant Problem.'" *Proceedings of the Tall Timbers Conference on Ecological Animal Control by Habitat Management* 7 (1978): 1–9.

Kraft, M. E. "U.S. Environmental Policy and Politics: From the 1960s to the 1990s." *Journal of Policy History* 12 (2000): 33–36.

Kuklick, H. "The Sociology of Knowledge: Retrospect and Prospect." *Annual Review of Sociology* 9 (1983): 287–310.

Landy, M., M. Roberts, and S. Thomas. *The Environmental Protection Agency: Asking the Wrong Questions*. New York: Oxford University Press, 1990.

Latour, B. *Science in Action: How to Follow Scientists and Engineers through Society*. Cambridge: Harvard University Press, 1987.

Lawrence, E. *Hunting the Wren: Transformation of Bird to Symbol*. Knoxville: University of Tennessee Press, 1997.

Lawrence, H. W. "The Geography of the U.S. Nursery Industry: Locational Change and Regional Specialization in the Production of Woody Ornamental Plants." Ph.D. diss., University of Oregon, 1985.

Lear, L. *Rachel Carson: Witness for Nature*. New York: W. W. Norton, 1997.

Lears, T. J. J. *No Place of Grace: Antimodernism and the Transformation of American Culture, 1880–1920*. New York: Pantheon Books, 1981.

Lennartz, F. E. "Modes of Dispersal of *Solenopsis invicta* from Brazil into the Continental United States–A Study in Spatial Diffusion." Master's thesis, University of Florida, 1973.

Leopold, A. *The River of the Mother of God and Other Essays*. Ed. S. L. Flader and J. B. Callicott. Madison: University of Wisconsin Press, 1991.

———. *A Sand County Almanac*. New York: Ballantine Books, [1949] 1970.

Lockwood, J. A. "Competing Values and Moral Imperatives: An Overview of Ethical Issues in Biological Control." *Agriculture and Human Values* 14 (1997): 205–10.

———. "The Ethics of Biological Control: Understanding the Moral Implications of Our Most Powerful Ecological Technology." *Agriculture and Human Values* 13 (1996): 2–19.

Löding, H. P. "An Ant." *U.S. Department of Agriculture Insect Pest Survey Bulletin* 9 (1929): 241

Lofgren, C. S., and R. K. Vander Meer. *Fire Ants and Leaf-Cutting Ants: Biology and Management*. Boulder: Westview, 1986.

Lofgren, C. S., W. A. Banks, and B. M. Glancey. "Biology and Control of Imported Fire Ants." *Annual Review of Entomology* 20 (1975): 1–30.

Lofgren, C. S., and D. E. Weidhaas. "On the Eradication of Imported Fire Ants: A Theoretical Appraisal." *Bulletin of the Entomological Society of America* 18 (1972): 17–20.

Lowi, T. J. *The End of Liberalism: Ideology, Policy, and the Crisis of Public Authority*. New York: Norton, 1969.

Maasen, S., E. Mendelsohn, and P. Weingart, eds. *Biology as Society, Society as Biology*. Dordrecht: Kluwer Academic Publishers, 1995.

MacIntyre, A. "Administrative Initiative and Theories of Implementation: Federal Pesticide Policy, 1970–1976." In *Public Policy and the Natural Environment*, ed. Helen M. Ingram and R. Kenneth Godwin, 205–38. Greenwich, Conn.: JAI Press, 1985.

Maeterlinck, M. *The Life of the Ant*. Trans. B. Miall. New York: John Day, 1930.

Marks, S. A. *Southern Hunting in Black and White: Nature, History, and Ritual in a Carolina Community*. Princeton: Princeton University Press, 1991.

McConnell, A. G. *Private Power and American Democracy*. New York: Knopf, 1966.

McEvoy, A. F. "Toward an Interactive Theory of Nature and Culture: Ecology, Production, and Cognition in the California Fishing Industry." *Environmental Review* 11 (1987): 289–305.

McKibben, B. *The End of Nature*. New York: Random House, 1989.

Mooney, H., and J. Drake, eds. *Ecology of Biological Invasions of North America and Hawaii*. New York: Springer-Verlag, 1986.

Moore, E. G. *The Agricultural Research Service*. New York: Praeger, 1967.

Morone, J. A. *The Democratic Wish: Popular Participation and the Limits of American Government*. New Haven: Yale University Press, [1990] 1998.

Morris, A. *South America*. New Jersey: Barnes and Noble Books, 1987.

Newsom, L. D. "Eradication of Plant Pests—Con." *Bulletin of the Entomological Society of America* 24 (1978): 35–40.

Palladino, P. *Entomology, Ecology and Agriculture: The Making of Careers in North America, 1885–1985*. The Netherlands: Harwood Academic Publishers, 1996.

———. "On 'Environmentalism': The Origins of Debates over Policy for Pest-Control Research in America, 1960–1975." In *Science and Nature: Essays in the History of the Environmental Sciences*, ed. M. Shortland, 201–24. Oxford: British Society for the History of Science, 1993.

Parrish J. R., and M. R. Potts. *The Great Science Fiction Pictures*. Metuchen, N.J.: Scarecrow Press, 1977.

Perkins, J. H. "Edward Fred Knipling's Sterile Male Technique for Control of the Screwworm Fly." *Environmental Review* 5 (1978): 19–37.

———. *Insects, Experts, and the Insecticide Crisis: The Quest for New Pest Management Strategies*. New York: Plenum Press, 1982.

Popham, W. L., and D. G. Hall. "Insect Eradication Programs." *Annual Review of Entomology* 3 (1958): 335–54.

Porter, S. D., and D. A. Savignano. "Invasion of Polygyne Fire Ants Decimates Native Ants and Disrupts Arthropod Community." *Ecology* 71 (1990): 2095–2106.

Porter, S. D., D. F. Williams, R. S. Patterson, and H. G. Fowler. "Intercontinental Differences in the Abundance of Solenopsis Fire Ants (Hymenoptera: Formicidae): Escape from Natural Enemies?" *Environmental Entomology* 26 (1997): 373–84.

Price, T. *The '84 Gamecocks: Fire Ants and Black Magic*. Columbia: University of South Carolina Press, 1985.

Quarles, J. *Cleaning up America: An Insider's View of the Environmental Protection Agency*. Boston: Houghton Mifflin, 1976.

Raudenbush, B. *Brilly and the Boot*. Conyers, Ga.: Panola Publisher, 1997.

Réaumur, R. A. F. *The Natural History of Ants*. Ed. W. M. Wheeler. New York: Knopf, 1926.

Redford, E. S. *Democracy in the Administrative State*. New York: Oxford University Press, 1969.

Rennie, Y. F. *The Argentine Republic*. Westport, Conn.: Greenwood Press, 1945.

Rhoades, R. B. *Medical Aspects of the Imported Fire Ant*. Gainesville: University Presses of Florida, 1977.

Rock, D. *Argentina 1516–1982: From Spanish Colonization to the Falklands War*. Berkeley: University of California Press, 1985.

Rome, A. *The Bulldozer in the Countryside: Suburban Sprawl and the Rise of American Environmentalism*. New York: Cambridge University Press, 2001.

Ross, K. G., and L. Keller. "Ecology and Evolution of Social Organization: Insights from Fire Ants and Other Highly Eusocial Insects." *Annual Review of Ecology and Systematics* 26 (1995): 631–56.

Rossiter, Margaret W. "The Organization of Agricultural Sciences." In *The Organization of Knowledge in Modern America*, ed. John Voss and Alexandra Oleson, 211–48. Baltimore: Johns Hopkins University Press, 1979.

Rudd, R. "The Irresponsible Poisoners." *The Nation* (30 May 1959): 496–97.

———. *Pesticides and the Living Landscape*. Madison: University of Wisconsin Press, 1964.

———. "Pesticides: The Real Peril." *The Nation* (28 November 1959): 399–401.

Russell, E.P. "Lost among the Parts Per Billion: Ecological Protection at the United States Environmental Protection Agency, 1970–1993." *Environmental History* 2 (1997): 29–51.

———. " 'Speaking of Annihilation': Mobilizing for War against Human and Insect Enemies, 1914–1945." *Journal of American History* 82 (1996): 1505–29.

———. "The Strange Career of DDT: Experts, Federal Capacity, and Environmentalism in World War II." *Technology and Culture* 40 (1999): 770–96.

———. *War and Nature: Fighting Humans and Insects with Chemicals from World War I to Silent Spring.* Cambridge: Cambridge University Press, 2001.

Saloutos, T. *The American Farmer and the New Deal.* Ames: Iowa State University Press, 1982.

Schapsmeier, E. L., and F. H. Schapsmeier. *Ezra Taft Benson and the Politics of Agriculture: The Eisenhower Years, 1953–1961.* Danville, Ill.: Interstate Printers and Publishers, 1975.

Schueler, D. G. *A Handmade Wilderness.* Boston: Houghton Mifflin, 1996.

Schulman, B. J. *From Cotton Belt to Sunbelt: Federal Policy, Economic Development, and the Transformation of the South, 1938–1980.* New York: Oxford University Press, 1991.

Scott, J. C. *Seeing Like a State: How Certain Schemes to Improve the Human Condition Have Failed.* New Haven: Yale University Press, 1998.

Scruggs, C. G. *The Peaceful Atom and the Deadly Fly.* Austin, Tex.: Jenkins Publishing, 1975.

Shanahan, E. W. *South America: An Economic and Regional Geography with an Historical Chapter.* London: Methuen, 1927.

Shapley, D. "Mirex and the Fire Ant: Declines in the Fortune of a 'Perfect' Pesticide." *Science* 172 (1971): 358–60.

Shores, E. F. "The Red Imported Fire Ant: Mythology and Public Policy, 1957–1992." *Arkansas Historical Quarterly* 53 (1994): 320–39.

Simmonds, I. G. *Environmental History: A Concise Introduction.* Oxford: Blackwell, 1993.

Sorenson, W. C. *Brethren of the Net: American Entomology, 1840–1880.* Tuscaloosa: University of Alabama Press, 1995.

Stephenson, C. "Leiningen versus the Ants." *Esquire* 10 (December 1938): 98–99, 235–41.

Sterling, W. L. "Imported Fire Ant . . . May Wear a Grey Hat." *Texas Agricultural Progress* 24 (1978): 19–20.

Stine, J. K. "Environmental Politics in the American South: The Fight over the Tennessee–Tombigbee Waterway." *Environmental History Review* 15 (1991): 1–24.

Stoddard, H. L. *The Bobwhite Quail: Its Habits, Preservation and Increase.* New York: Scribner, 1931.

———. *The Cooperative Quail Study Association: May 11, 1931–May 1, 1943.* Tall Timbers Research Station, Miscellaneous Publications, no. 1. Tallahassee, Fla.: Tall Timbers Research Station, 1961.

Suarez, A. V., D. A. Holway, and T. J. Case. "Patterns of Spread in Biological Invasions Dominated by Long-Distance Jump Dispersal: Insights from Argentine Ants." *Proceedings of the National Academy of Sciences* 98 (2001): 1097.

Taber, S. W. *Fire Ants.* College Station: Texas A&M University, 2000.

Tenner, E. *Why Things Bite Back: Technology and the Revenge of Unintended Consequences.* New York: Knopf, 1996.

Terborgh, J. *Requiem for Nature.* Washington, D.C.: Island Press, 1999.

Thomason, M. V. R., ed. *Mobile: The New History of Alabama's First City.* Tuscaloosa: University of Alabama Press, 2001.

Throup, H. C. "Clarence Cottam: Conservationist: The Welder Years." Ph.D. diss., Brigham Young University, 1983.

Trager, James C. "A Revision of the Fire Ants *Solenopsis geminata* Group (Hymenoptera: Formicidae: Myrmicinae)." *Journal of the New York Entomological Society* 99 (1991): 141–98.

Tschinkel, W. R. "Sociometry and Sociogenesis of Colonies of the Fire Ant *Solenopsis invicta* during One Annual Cycle." *Ecological Monographs* 64 (1993): 425–57.

Twelve Southerners. *I'll Take My Stand: The South and the Agrarian Tradition*. New York: Harper and Brothers, 1930.

Ullman, E. L. "Mobile: Industrial Seaport and Trade Center." Ph.D. diss., University of Chicago, 1942.

Van den Bosch, R. *The Pesticide Conspiracy*. Garden City, N.Y.: Doubleday, 1978.

Vander Meer, R. K., K. Jaffe, and A. Cedeno, eds. *Applied Myrmecology: A World Perspective*. Boulder: Westview, 1990.

Vinson, S. B. "Invasion of the Red Imported Fire Ant (Hymenoptera: Formicidae): Spread, Biology, and Impact." *Bulletin of the Entomological Society of America* 43 (1997): 23–39.

Vogel, S. *Against Nature: The Concept of Nature in Critical Theory*. Albany: State University of New York Press, 1996.

Warren, B. *Keep Watching the Skies: American Science Fiction Movies of the Fifties*. Vol. 1. London: McFarland, 1982.

Wellford, H. *Sowing the Wind*. New York: Grossman Publishers, 1972.

Wellock, T. R. *Critical Masses: Opposition to Nuclear Power in California, 1958–1978*. Madison: University of Wisconsin Press, 1998.

Wernick, S. *The Fire Ants*. New York: Award Books, 1976.

Wheeler, W. M. "Notes about Ants and Their Resemblance to Man." *National Geographic* 23 (1912): 731–66.

———. "Notes on the Brazilian Fire Ant *Solenopsis saevissima* F. Smith." *Psyche* 23 (1916): 142–43.

———. *The Social Insects: Their Origin and Evolution*. London: Keegan Paul, Trench, Trubner, 1928.

———. *Social Life among the Insects*. New York: Harcourt, Brace, 1923.

White, R. "American Environmental History: The Development of a New Historical Field." *Pacific History Review* 54 (1985): 297–335.

———. "Environmental History, Ecology, and Meaning." *Journal of American History* 76 (1990): 1111–16.

———. *The Organic Machine: The Remaking of the Columbia River*. New York: Hill and Wang, 1995.

Whitten, J. L. *That We May Live*. Princeton, N.J.: Van Nostrand Press, 1966.

Wilcove, D. S., M. J. Bean, R. Bonnie, and M. McMillan. "Rebuilding the Ark: Toward a More Effective Endangered Species Act for Private Land." 5 December 1996, available at www.environmentaldefense.org/documents/483_Rebuilding%20the%20Ark, accessed 17 October 2002.

Wilcox, R. W. " 'The Law of Least Effort': Cattle Ranching and the Environment in the Savanna of Mato Grosso, Brazil, 1900–1980." *Environmental History* 4 (1999): 338–68.

Williams, R. *Problems in Materialism and Culture*. London: Verso, 1980.

Wills, G. *A Necessary Evil: A History of American Distrust of Government*. New York: Simon and Schuster, 1999.

Wilson, E. O. *The Diversity of Life*. Cambridge: Harvard University Press, 1992.

———. "The Evolutionary Significance of the Social Insects." In *Insects, Science and Society*, ed. D. Pimentel, 25–31. New York: Academic Press, 1975.

———. "The Fire Ant." *Scientific American* 198 (March 1958): 36–41, 160.

———. *The Insect Societies*. Cambridge: Harvard University Press, 1971.

———. "In the Queendom of Ants: A Brief Autobiography." In *Leaders in the Study of Animal Behavior: Autobiographical Perspectives*, ed. Donald A. Dewsbury, 465–86. Lewisburg, Pa: Bucknell University Press, 1985.

———. "Invader of the South." *Natural History* 68 (1959): 276–81.

———. *Naturalist.* Washington, D.C.: Island Press, 1994.

———. "O Complexo *Solenopsis saevissima* na America do Sul (Hymenoptera: Formicidae)." *Memorio do Instituto Oswald Cruz* 50 (1952): 49–68.

———. "Origin of the Variation in the Imported Fire Ant." *Evolution* 7 (1953): 68–79.

———. *Success and Dominance in Ecosystems: The Case of the Social Insects.* Germany: Ecology Institute, 1990.

———. "Variation and Adaptation in the Imported Fire Ant." *Evolution* 5 (1951): 68–79.

Wilson, E. O., and W. L. Brown. "Recent Changes in the Introduced Population of the Fire Ant *Solenopsis saevissima* (Fr. Smith)." *Evolution* 12 (1958): 211–18.

Winston, M. *Nature Wars: People vs. Pests.* Cambridge: Harvard University Press, 1997.

Woodward, C. V. *The Burden of Southern History.* Baton Rouge: Louisiana State University Press, 1993.

Worrell, A. C. "Pests, Pesticides, and People." *American Forests* 66 (1960): 39–81.

Worster, D. "History as Natural History: An Essay on Theory and Method." *Pacific History Review* 53 (1984): 1–19.

———. "Nature and the Disorder of History." *Environmental History Review* 18 (1994): 1–15

———. *Nature's Economy: A History of Ecological Ideas.* New York: Cambridge University Press, [1977] 1996.

———. "Transformations of the Earth: Toward an Agro-Ecological Perspective in History." *Journal of American History* 76 (1990): 1087–1106.

Index